Artificial Rain Production by Endothermic Reactions

(By Laser Systems as Similar Natural Lightning Phenomena in the Atmosphere)

"Green revolution in the whole world for all human being".

Innovative Rainmaking Technology is scientifically and practically proven in laboratory cloud chambers, as well as in the atmosphere. This includes as Ref. "Laser-induced condensation and formation of waterdrops in laboratory cloud chamber as well as in the atmosphere up to 75m altitude"

Innovative Rainmaking Research Association, India (IRRA Scientist),

Shivshankar K. Chopkar, D. K. Chakrabarty,

Sewagram- 442102, Dhanvantari Nager-30, Wardha, Maharashtra, India. E- skc.arr@rediffmail.com ,E- irra.scientistgroup@gmail.com, India, Mob. +91 9420445108, Website-www.irraindia.org.

BLUEROSE PUBLISHERS
India | U.K.

For permissions requests or inquiries regarding this publication,
please contact:

BLUEROSE PUBLISHERS
www.BlueRoseONE.com
info@bluerosepublishers.com
+91 8882 898 898
+4407342408967

ISBN: 978-93-5989-562-8

First Edition: May 2024

Shri Sadguru Parmhans Ramchandra Maharaj Namo Namha"

"No problem in whole life"

"No Problem"

For each and everywhere!

There are so many problems!

But no problem!

If, you are passenger….

And, God is driver!

Then, No problem in whole life!

Otherwise, each and everywhere!

There, are so many problems!

But, No problem!

If, you are a passenger!

And God is driver!

Then, No problems in whole life!

By

Shivshankar chopkar

(Devotees of shri Ramchandra's God)

Dedication

Innovative Rainmaking Technology is scientifically and practically proven in laboratory cloud chambers, as well as in the atmosphere. This includes as "Laser-induced condensation and formation of water drops in laboratory cloud chamber as well as in the atmosphere up to 75m altitude.

"Artificial rainmaking by using high power laser which initiates endothermic reactions, similar to nature's lightning phenomenon, onboard Aircraft in the atmosphere"

Acknowledgement

We express our sincere thanks to scientists: Prof D.K. Chakarbarty (PRL Ahmadabad), Prof Mamata R. Lanjewar (RTM University Nagpur), Prof A. P. Deshpande (Principal, Science College), Prof. Padmanabhan Murthy (J.N. University, New Delhi), Prof Umesh Kulshetra (J.N. University, New Delhi), Prof K. S. Korgaokar (Pune University, Pune), Dr. A.L. Agarwal (NEERI Nagpur), Dr. Nitin Saraf (B.D. Engineering College, Sewagram), Dr. K. M. Kharate, (GM Engineering College & Research Centre, Sheogao), Dr. Mrs. Sumanlatha Pandey, Mr. Shyam Ujjaninkar (Engineer), Mr. Rajesh Iyengar (P. N. College, Nanded), Mr. Anant Kutemate (Manager), for their valuable suggestions and guidance for project proposal.

(IRRA Scientists research activities with rainmaking project proposal for Demonstration)

"Innovative rainmaking Technology used for Artificial Rain on large scale by Laser systems as similar natural Lightning phenomena in the atmosphere"

Abstract: Challenging research activities by IRRA Scientists have developed an "Innovative Rainmaking Technology " which is scientifically and practically proven in Laboratory cloud chambers and the atmosphere up to 75m altitude as Ref. "Laser induces condensation and water drops formation in laboratory clouds chamber & atmosphere".

Now IRRA Scientists have developed two research project proposals as below.....

1) "Artificial rainmaking by Peta Watt (10^{15}Watt) double core Laser system from ground initiation of Endothermic Reactions, as similar natural lightning phenomena in the atmosphere". It covers ~16 Km2, an area... Estimate cost 9.02 Cr in Indian Rs/ USA 0.72 million Dollar)

2) "Artificial rainmaking by using high power terawatt (10^{12} Watt) mobile Laser Which initiates endothermic reactions, as a similar natural lightning phenomenon, onboard Aircraft with multiple lightning in the atmosphere". It covers ~450 Km2, an area...& estimates cost 230 Cr./USA dollar 28.75 million.

IRRA scientists have prepared a new project proposal with a design, budget & estimate, work plan etc. on "Artificial rainmaking by laser system which initiate endothermic reactions, as similar natural lightning phenomenon, onboard Aircraft & Ground in the atmosphere" which is helpful for demonstrations in the atmosphere, enclosed here! IRRA Scientists would like to collaborate/partner with any Government of the country, Environment Science and Technology Department / Science and technology Institution/Foundation /Company etc. for a Demonstration of "Innovative rainmaking Technology used for Artificial Rain on large scale by Laser systems as similar natural Lightning phenomena on board Aircraft in the atmosphere" for three years, measuring & fixing atmospheric parameters as ex.- temperature, pressure, humidity etc. for maximum rainmaking in the atmosphere with project proposal estimates cost 230 Cr./USA dollar 28.75 million

Contents

1.0: "Beginning Of A Research On Artificial Rain Based On Natural Process And Its Subsequent Achievement" By Goddess Sonai My Mother At An Early Age For The Welfare Of All Living Beings: - 13

2.0: Achievements By IRRA Scientists: 15

3.0: Mission, Vision, Objects: 17

4.0: Flow Chart For Rainmaking: 18

5.0 "National And International Patent File" 22

6.0: "Request For Help" 22

7.0: "A Challenging Research Work On "Innovative Rainmaking Technology By Laser" 23

8.0: "Research Activities By IRRA Scientist Group" 25

9.0 Now IRRA Scientists have developed two research project proposals as below 25

10.0: Two Project Proposal on Rainmaking Technology for Demonstration from Ground & onboard Aircraft in the atmosphere 27

10.11:Figures For Demonstration On Rainmaking: 69

10.12 :Budget For Rainmaking Project Proposal For Three Years In The Atmosphere: 70

10.13 : Work-Plan For Three Years: 74

14.0: Request To Encourage IRRA Scientists Research Activities: 76

15.0: "Request Letter" 77

16.0: Research Paper Publish In National And International Journal By IRRA Scientist Group" 82

17.0: Research Activities And Achievement On Rainmaking Technology By IRRA Scientist 87

18.0: "Achievements By IRRA Scientist"..88

19.0: Registration Certificate For IRRA 95

20.0: International Scientists Award..97

(Innovative Rainmaking Technology)

1.0: "Beginning of A Research on Artificial Rain Based on Natural Process and Its Subsequent Achievement" By Goddess Sonai My Mother at An Early Age for The Welfare of All Living Beings: -

This is a dream for bringing green revolution in the entire world. Let's make this dream come true. The work of bringing a green revolution in the universe has been entrusted to us by God to make all living beings happy. With a hearty thanks to the Almighty God, we will follow the footsteps of God and complete this project on artificial rain research based on natural process. This research shall also bring to the fourth a technology which can halt excessive rains whenever there is a cloud burst. This means we can have rains as much as we need and wherever we want it to happen. God will carry out the work assigned to us.

This happened somewhere during the early 70s. From the year 1972 to 1976, there was no rain for three to four consecutive years. Drinking water was not available. There was a huge shortage of food and water. People started going for labor work under the employment guarantee schemes. During the course of my work with my mother, I got a blister on my palm which broke out. One day when my mother had gone two kilometers to fetch drinking water, she left a tiffin containing red sorghum bread, onion and red chutney. I opened the tiffin and started eating the food items.

My mother later arrived and brought some water. When my mother saw me eating, she got infuriated. Seeing my sad face, she then asked me to eat and I rushed and started gulping the food. But my mother told me to wash my hands. I just washed the tips of my fingers and continued eating. My mother upbraided me for not washing my hands properly and poured the cold water from the well freshly brought by her. The blistered hand started hurting me like hell. I cried. My mother looked at my hand. The

(1972) In my child life, my mother Smt. Sonai Chopkar give me 'Guru Mantra' that

"If you do good to the world, it ultimately became your good"

मा सोनाई का गुरु मंत्र "दुसरो के कल्याण में ही अपना कल्याण है"

hand was sore. The skin was peeled off. There were tears in her eyes. My mother cursed God. In common parlance she cussed the God "Oh! You wretched God, you have no wisdom. Why don't you pour rains?" A thought started circling my mind. Had it rained why did I have to come to work? Mother burst into tears. I too started weeping. In the scorching sun with no shade of the trees I was standing there helpless. A lot of people came to work as if there was some village fair. We were eating chutney and bread under the blistering sun, under the shadow of an imaginary tree. Tears started rolling on my cheeks. A thought started forming in my mind. "Why didn't it rain? Why didn't it rain?" I was thinking a lot. There was no second thought in my mind. Drops of sweat were falling on the ground due to intense heat from my body. After finishing work, we went home. Even while returning home there was a huge commotion of thoughts in mind. Why didn't it rain? What does it need to make rain fall?

As usual, we came to the temple in the evening to pay obeisance to God. Now I had a resolution in my mind. I will have to undertake lifelong research to make this dream come true. With Goddess Sonai's my mother blessing, the work on making this dream reality started working. It is still continuing. This thought has been instilled in our minds by Goddess Sonai my mother at an early age for the welfare of all living beings by providing artificial rain based on natural process. Secondly, it should rain whenever we need and wherever we want. No doubt it has become our "Guru Mantra" that 'if you do good to the world, it ultimately becomes your own good". Fortunately, this Guru Mantra has reached its culmination by the grace of Sadguru Parmhanse Ramchandra Maharaj.The research work started from 1975. The first research paper was published as "Effect of endothermic reactions associated with lightning on atmospheric chemistry" in Indian Journal of Radio & Space Physics Vol. 22, April 1993, pp123-131, under the aegis of Atmospheric Rainmaking Research Society, Chimur Dist. Chandrapur, registered under Maharashtra Government in 1998 as NGO. "Artificial rainmaking system in a way of natural phenomena" research paper was also published in "Indian Journal of Science and Technology", www.indjst.org, Vol.1No.6 (Nov.2008). IRRA Scientist Group also published research papers in national and international Journal as Fellow as stated below.

2.0: Achievements By IRRA Scientists:

- *IRRA Scientist Group Member (Dr. S.K. Chopkar) presented research paper as "Artificial rainmaking by Laser system similar to natural lightning phenomena in the atmosphere" in "NATIONAL WORKSHOP ON ATMOSPHERIC CHEMISTRY (NWAC-99), Indian Institute of Tropical Meteorology, Pune, Maharashtra.*
- *IRRA Scientist Group Member (Prof. D.K. Chakrabarty) presented research paper as "Femtosecond terawatt laser system to produce artificial rain" in International Conference, Institute of Radio Physics & Electronics, University of Calcutta,92 A P C Road, Kolkata 700 009, India.*
- *IRRA Scientist Group Member (Dr. S.K. Chopkar) presented research paper as "Novel technology for artificial rainmaking by laser system similar to natural lightning phenomena" in "International Environment Sahara Conference at Dakhala, Morocco, 27-28 September 2018".*
- *IRRA Scientist Group has been awarded Outstanding Scientist Award, VDGOOD, 14-15, Sep.2019, Chennai, India.*

Our Mission is "Even Distribution of Rain for bringing about Green Revolution in the Globe" and our Vision is "Global Prosperity through Artificial Rainmaking in the whole world for all the human beings".

Our Main objectives are: -

a) Natural Seeding for Artificial Rainmaking……

b) To control excess rainfall with well distribute of rainmaking…..

Our novel technology is awarded with International Publication Patent No: WO/2008/062441 and Revised International Patent Application No: PCT/IN2017/000105.

International Australian Government, IP Australia Certificate ofGrants Innovation Patent number: 2020101897 (2021). Indian Patent Number: 238000 and Revised Indian Patent File Number: 201721008920.

Our research papers have been published in national and international Journals such as "The International Journal of Meteorology, Volume 35, and number 355 on November 2010 (www.ijmet.org). For more details there can be a Google search on as S.K. Chopkar or "Artificial rainmaking by endothermic reactions" or www.irraindia.org can be visited. Our aim is "Green revolution in the whole world for all human being".

IRRA is a private organization which works under "NO PROFIT AND NO LOSS" basis. We work for the welfare of the society. We have no source of income and no funding from anyone. Our scientists work here on voluntary basis during their free time. We manage by contributing money from our own resources.

Now, Innovative Rainmaking Research Association is registered with Government of India, Ministry of Corporate Affairs under Section-8, on a "no profit, no loss basis". We have developed "Innovative rainmaking technology by laser system initiation endothermic reactions similar to natural lightning phenomena in the atmosphere"

A laser beam has to be shot into the cloud region of the atmosphere to create high temperature. This high temperature will break the bonds of atmospheric N_2 and O_2 and produce N and O which will be in excited state (N, O*). These excited N* and O* are very unstable. They immediately fall to the ground through endothermic reactions.*

These endothermic reactions absorb heat from the cloud region. As a result, temperature of the cloud region falls, *condensation takes place and rain drops are formed which act as natural seeds ultimately resulting in rains. It is used as rain drain by Laser system from ground for creating green environment. "Laser photons photo-dissociate atmospheric compounds N_2 and O_2 and form ozone (O_3) and nitrogen molecules (NO). Increase of O_3 and NO concentration after lightning has also been experimentally observed. This lightning phenomenon created through artificial lightning by laser system can produce rain in the atmosphere. This has been practically proved as "Laser-induced water condensation in air". Scientists have succeeded in obtaining raindrops from an altitude of 45 to 75m of the atmosphere by terawatt mobile laser.*

For "Laser-induced condensation and water drops formation in laboratory cloud chamber" to happen only endothermic reactions are responsible for condensation and water drops formation. Ozone (O_3) and nitrogen molecules (NO) are measured after Laser shooting in cloud chamber. These NO and O_3 formation are endothermic reactions, these reactions require large amount of heat energy which is taken from surrounding atmospheric clouds, and condensation takes place for formation of water drops. Our phenomena "Innovative rainmaking technology" practically proved.

IRRA Scientist Group proposes laser system of specification: 10^{12}watt, 800nm, 500mJ, 120fs and 10Hz for this research project. The results could be a great game changer and of immense benefit to human being for bringing up a "Green revolution in the whole world for all human being".

3.0: Mission, Vision, Objects:

IRRA Scientist Group, proposed "Novel technology for Rain Enhancement using high power Laser system initiation endothermic reactions.

"Laser-induced condensation & water drops formation" Process has been practically proved in the laboratory cloud's chamber as "production of ozone and nitrogen oxides by laser". O3 & NO are endothermic reactions which responsible for condensation and water drops formation, these rain drops act as natural seeding,

enhancing rain in a way of natural lighting phenomena. Our many research papers have been published in national and international Journals; also, Innovative *Rainmaking* Technology has been Patent file for national and international PCT publish. Our aim is "Green revolution in the whole world for all human being. This is God Gift for us. Search on Google as S.K. Chopkar or *"Rainmaking* by endothermic reactions" or website *-http://www.irraindia.org* you know more details about "Innovative Rainmaking Technology", it's scientifically & practically proved in Laboratory cloud's chamber & as in the atmosphere.

"MISSION"

"Green Revolution in the whole world for all human being"

"Vision"

"Global Prosperity through Artificial Rainmaking".

" Objectives"

1) "Novel technology for artificial rainmaking using high power Laser systemin the atmospheric clouds initiation endothermic reactions, in a way of natural lightning phenomena"

2) "Well distribution of rainmaking in all over" also rain drain in Dam area

4.0: Flow Chart for Rainmaking:

(Flow Chart)

"Innovative Rainmaking Technology by Laser System initiating Endothermic Reactions similar to Natural Lightning Phenomenafor Rainmaking in the Atmosphere"

↓

** We well know, after lightning, precipitation is formed and, heavy rainfall occurs. This Lightning phenomenon for Rainmaking is caused by endothermic reactions as......*

↓

**Lightning creates high temperature, at high temperature, bondsofN2 (74%) and O2 (23%) break out into excited N* and O* as....,*

↓

$$N_2 : N \equiv N \rightarrow N^* + N$$

$$O_2: O = O \rightarrow O^* + O$$

↓

**Excited N* and O* are unstable, react with each other and NOand O_3 are formed.*

↓

$$N^* + O_2 \rightarrow NO \quad + O - \Delta H\, (\Delta H = 43.2 kcal/mol)$$

$$O^* + O_2 + M \rightarrow O3 + M - \Delta H\ (\Delta H = 67.7 kcal/mol)$$

↓

** NO and O_3 formations are endothermic reactions. Heat energyrequired for these reactions is taken from the surrounding atmospheric clouds.*

↓

**As a result, these reactions will produce, temperature falls, andcondensation takes place, which is basic need for rain drops formation, these rain drops act as natural seeding, rain is in analogous way similar to rains created in nature by lightning.*

↓

** Increase of O_3 and NO concentration after lightning has alsobeen experimentally observed in the atmosphere.*

↓

** Natural lighting phenomena has been practically demonstratedin the laboratory cloud chamber by high power Laser shooting,as result "Laser-induced condensation & water drops formation" is observed.*

↓

** Process has been practically proved in the laboratory as "production of O3 (Ozone) and NO (Nitrogen Oxides) by lasershooting in cloud chamber".*

↓

**Laser photons photo-dissociate atmospheric compounds N_2 andO_2 and from O3 and NO which are endothermic reactions, required lot of heat energy, absorbs from surrounding atmospheric clouds, condensation & precipitation formed enhance rain, in the atmosphere.*

↓

** Lightning phenomenon created through artificial lightning by laser system can produce rain in the atmosphere has been practically proved as "Laser-induced water condensation in air".*

**Scientists have succeeded in obtaining raindrops from an altitude of 75m of the atmosphere by terawatt mobile laser.*

↓

**IRRA Scientist Group proposes laser system of specification of 10^{12} watt, 800nm, 500mJ, 120fs and 10Hz for this research project to create artificial lightning which initiates endothermicreactions, similar to Natural Lightning Phenomena for Rainmaking in the atmosphere.*

↓

In Laboratory cloud chamber, formation of these tiny water dropsparticles are converted into rain drops by acceleration and turbulence due to wind force in nature and Rainmaking in the atmosphere.

↓

**Innovative Rainmaking technology is practically proved and successfully experimented and is known as "Laser makes rain".*

** Process is economical, harmless and eco-friendly and can be switched on and off by remote control from the ground. It is mostuseful for human beings particularly for farmers.*

↓

**Please search on Google as "Rainmaking by Endothermic reactions "or visit our web site http://www.irraindia.org*

↓

**Our aim is "Production of Rainmaking when and where requiredfor bringing about a green revolution in the whole world for the humankind".*

↓

**If our finding could be used to help scientists create new alternative method for Rainmaking, the resulting techniques could be of inestimable value.*

↓

Additional uses of Innovative Rainmaking Technology for raindrain in Dam catchment area.

↓

Water & Food are the basic needs of all human beings as well asanimals, trees, etc. which depends on rain; it is easily solved by artificial rain making.

A group of European scientists working on artificial rain said in 2010 "Firing extremely powerful laser pulses through humid air can stimulate the formation of clouds, according a team of European scientists. They say that the effectiveness of this method is much easier to gauge than traditional cloud- seeding techniques and that it could provide a practical means of triggering rainfall". (Search on Google as Laser makes rain heavily 2010).

"Novel Rainmaking Technology scientifically & practically proven in the atmosphere"

5.0 "National and International Patent File"

Following are the two patents on this topic.

** International Australian Government, IP Australia Certificate of Grants , Innovation Patent number: 2020101897 (2021).*

**International Publication Patent No: -WO/2008/062441 (2007) and Revised International Application No.PCT/IN2017/000105 (2017).*

**Indian Patent Number: - 238000 (2007) and Revised Indian Patent FileNumber 201721008920 (2017).*

6.0: "Request for Help"

Now days, IRRA scientist group are struggling with funds for research activities. Your little help may be most useful for "Green revolution in the whole world for all human being". Your co-operation is solicited. Innovative Rainmaking Research Association, Bank A/C (80G and 12A).

With best regards

Yours faithfully

IRRA Scientist Group

Bank Account Details

1) *Bank Name-CENTRAL BANK OF INDIA, MIDC SEWAGRAM ROAD--442102, Warud, Dist.-Wardha. (Maharashtra.), India.*

Bank A/c Name: - Innovative Rainmaking Research Association, (80G and 12A) Sewagram, Dhanvantri Nager-30, Bank A/c No. 3806623156, IFSC Code No.: - CBIN0282189, Bank A/C (80G and 12A).

2) *Bank Name-CENTRAL BANK OF INDIA, SEWAGRAM BRANCH--442102, Warud, Dist.-Wardha. (Maharashtra.), India.*

Bank A/c Name: - International Rainmaking Research Academy,

Sevagram, Dhanvantri Nager-30,

Bank A/c No. 3841144252, IFSC Code No.: - CBIN0280697

With best regards

BY - IRRA Scientist Group.

INNOVATIVE RAINMAKING RESEARCH ASSOCIATION (80G and 12A)

*Shivshankar Kanhuji Chopkar

Sevagram, Dhanvantri Nagar-30, Varud, Dist.- Wardha-442102, Maharashtra, India.

Cell +91 9420445108, EMAIL: - skc.arr@rediffmail.com / irra.scientistgroup@gmail.com

Website -http://www.irraindia.org

Innovative Rainmaking Research Association, Bank A/C (80G and 12A)

7.0: "A Challenging Research Work On "Innovative Rainmaking Technology by Laser"

"Innovative Rainmaking Technology" is scientifically and practically proven in Laboratory clouds chamber as well as in the atmosphere. As "Laser induce condensation and water drops formation in Laboratory clouds chamber as well as in the atmosphere.

A challenging research project on "Innovative rainmaking technology by laser which is similar to natural lightning phenomena from the ground level". A number of scientific and practical tests have been conducted in cloud chambers and in the atmosphere to prove this hypothesis. The use of innovative rainmaking technology is most beneficial for the "Green Revolution in the whole world for all humankind".

Scientists of IRRA Group are involved in rainmaking by laser techniques anytime, anywhere, as per human needs for rainfall in the atmosphere

with more than 65% humidity. Nature has given them as the skill for a "Green revolution in the whole world for all human beings".

The IRRA scientist Group has already demonstrated "Innovative Rainmaking Technology" in the Laboratory cloud's chamber successfully, and results have been published in "Indian Journal of Science & Technology", http://indjst.org, vol.1, No.6 (2008).IRRA scientists is involved for the last 37 years in the field of research for success on "Innovative Rainmaking Technology".

Now, the project proposal, the design of the laser system, the Budget estimate, and the work plan are ready with IRRA Scientist Group, India. IRRA Scientist is ready for demonstration and collaboration with the Government for funding purposes for project proposal on "An Innovative Rainmaking Technology using Laser system from ground to form raindrops acting as the natural seeding process due to initiation of Endothermic Reactions, for rain enhancement in large scale in the atmosphere", project proposal enclosed herewith for your kind consideration I hereby strongly recommend for Nomination of IRRA Scientist Group from India for Excellence Innovative Research work in the Field of Rainmaking Technology for your valuable environment Award.

IRRA Scientist Group, India

INNOVATIVE RAINMAKING RESEARCH ASSOCIATION

**Shivshankar Kanhuji Chopkar*

Sevagram, Dhanvantri Nagar-30, Varud, Dist.- Wardha-442102, Maharashtra, India.

Cell +91 9420445108,

EMAIL: - skc.arr@rediffmail.com / irra.scientistgroup@gmail.com

Website -http://www.irraindia.org

8.0: "Research Activities by IRRA Scientist Group"

A challenging research project on "Innovative rainmaking technology by laser, which is similar to natural lightning fromthe ground level" A number of scientific and practical tests have been conducted in cloud chambers and in the atmosphere to prove this hypothesis. The use of innovative rainmaking technology is most beneficial for the "Green Revolution in the whole world for all humankind" Scientists at IRRA aim for rainmaking by laser system anytime, anywhere, as per human needs for rainfall in the atmosphere with available more than 65% humidity. God has given us this as a gift.

Already, the IRRA scientist Group has demonstrated "Innovative Rainmaking Technology" in the Laboratory cloud's chamber successful, and results have been published in "Indian's Journal of

Science and Technology", http://indjst.org,vol.1, No.6.

Now, the project proposal, the design of the laser system, the bugged estimate, and the work plan are ready with IRRA Scientist Group, India. IRRA Scientist is ready for demonstration and collaboration with the Government for funding purposes for project proposal on "An Innovative Rainmaking Technology using Laser system from ground to form raindrops acting as the natural seeding process due to initiation of Endothermic

Reactions, for rain enhancement in large scale in the atmosphere", project proposal with us for your kind consideration.

Innovative Rainmaking Technology is scientifically and practically proven in laboratory cloud chambers, as well as in the atmosphere. This includes as "Laser-induced condensation and formation of water drops in laboratory cloud chamber as well as in the atmosphere up to 75m altitude.

"Artificial rainmaking by using high power laser which initiates endothermic reactions, similar to nature's lightning phenomenon, onboard Aircraft in the atmosphere"

Innovative Rainmaking Research Association (IRRA Scientist), India

Shivshankar K. Chopkar, D. K. Chakrabarty, Nitin Saraf, D. Sable, M.R. Lanjewar, K.R. Gangakhedkar, Shyam Ujjainkar, R.M. Kharate, A.P. Deshpande, Umesh Kulshetra. Prof. Dr. Jzsef Dr. Steier

Sewagram- 442102, Dhanvantari Nager-30, Warud, Dist-. Wardha, Maharashtra, India.

E-skc.arr@rediffmail.com ,E- irra.scientistgroup@gmail,com, Mob.+91 9420445108,Website-www.irraindia.org.

9.0 Now IRRA Scientists have developed two research project proposals as below.....

1) "Artificial rainmaking by Peta Watt (10^{15}Watt) double core Laser system from ground initiation of Endothermic Reactions, as similar natural lightning phenomena in the atmosphere". It covers ~16 Km2, an area... Estimate cost 9.02 Cr in Indian Rs/ USA 0.72 million Dollar)

2) "Artificial rainmaking by using high power terawatt (10^{12} Watt) mobile Laser Which initiates endothermic reactions, as a similar natural lightning phenomenon, onboard Aircraft with multiple lightning in the atmosphere". It covers ~450 Km2, an area...& estimates cost 230 Cr./USA dollar 28.75 million.

IRRA scientists have prepared a new project proposal with a design, budget & estimate, work plan etc. on "Artificial rainmaking by laser system which initiate endothermic reactions, as similar natural lightning phenomenon, onboard Aircraft & Ground in the atmosphere" which is helpful for demonstrations in the atmosphere, enclosed here! IRRA Scientists would like to collaborate/partner with any Government of the country,

10.0 "Two Project Proposal on Rainmaking Technology for Demonstration from Ground & onboard Aircraft in the atmosphere".

(10.1 Project proposal For Rainmaking from Ground)

"Artificial rainmaking by Peta Watt (10^{15} Watt) double core Laser system from ground initiation of Endothermic Reactions, as similar natural lightning phenomena in the atmosphere"

S. K. Chopkar, D. K. Chakrabarty, K. R. Gangakhedkar, Umesh Kulshetra. Prof. Dr. Jzsef Dr. Steier

Sewagram- 442102, Dhanvantari Nager-30, Wardha, Maharashtra, India.

E-skc.arr@rediffmail.com, E- irra.scientistgroup@gmail.com, Website-www.irraindia.org

Abstract

In the atmosphere, after lightning, precipitation is formed and heavy rainfall occurs. This is a well-known process. This natural lighting phenomenon has been practically demonstrated in the laboratory cloud chamber, as ***"Laser-induced condensation & water drops formation"*** *and* ***"Water drops formation by each and every laser shot in the cloud chamber".*** *In this process, lightning/laser creates high temperature which breaks the bonds of N_2 and O_2 to form excited N^* and excited O^*. Total heat energy of lightning/laser is completely utilized for breaking the bonds of N_2 and O_2 (Chopkar, 1993). These excited N^* and excited O^* move to new place by wind and undergo reactions to form NO and O_3 which are endothermic reactions. Heat energy required for these reactions are taken from the surrounding atmospheric clouds. As a result of these reactions, temperature falls, condensation takes place, rain drops form, these raindrops act as natural seeding process, to form another sets of rain drops, seeds are created and rain occurs in analogous way similar to rains created in nature by lightning. In this process, white clouds convert into black rainy clouds for rainmaking in the atmosphere. This process has been practically proved in the laboratory as* ***"Production of ozone and nitrogen oxides by laser filamentation".*** *It is believed that "Laser photons photo-dissociate atmospheric compounds N_2 and O_2 form ozone (O_3) and nitrogen molecules (NO).* ***"Increase of O_3 and NO concentration after lightning has also been experimentally observed".*** *This lightning phenomenon created through artificial lightning by Peta Watt double laser pulse system can produce rain in the atmosphere has been practically proved as* ***"Laser-induced water condensation in air".*** *Scientists have succeeded in obtaining raindrops from an altitude of 75m of the atmosphere by terawatt mobile laser.*

"IRRA Scientist Group propose Peta Watt (10^{15}Watt) double core laser system from ground of specification: 10^{15}watt, 800nm, 500mJ, 120fs and 10Hz for this research project". *The results could be of immense benefit to human being.*

Keywords: Artificial rain; Atmospheric cloud; Condensation; Endothermic reactions; Laser system; Precipitation; Raindrops; Rainfall; Natural Seeding.

10.1: Introduction:

Several attempts have been made by various researchers to create artificial rain by laser. Golde (1977) from a number of radar observations has reported that intense precipitation is not even present in the clouds before the first discharge but develops abruptly in the same region after discharge from which the lightning flashes originate. Carls and Brock (1987) heated the atmosphere by a laser pulse up to 1600 to 2800K and observed water droplet formation. They predicted that high temperature causes ionization of N_2 and O_2 and, when this ionized air is subjected to more radiation, avalanche breakdown of air can occur. Braun et al. (1995) have observed laser induced condensation and water drops formation by shooting self-channeling of high- peak power femtosecond laser pulses in the air. Yoshihara et al. (2007) have shown that the pulsed UV-laser irradiation of ambient air induces formation of water droplets or small ice particles in the laboratory. They also observed that [O] formed in this process quickly reacts with O_2 molecules to form O_3. Rohwetter et al. (2010) have shown that ionized filament, generated by ultra-short wave laser pulses induce water-cloud condensation in the sub-saturated atmosphere in the altitude region between 45m and 75m. A team, called terra-mobile-group (TMG), consisting of scientists from Switzerland, Germany and France, have been trying to create artificial rain by laser (Kasparian et al.2000; 2003; Mejean et al. 2006; Rohetteret al. 2010; Kasparian et al. 2012). They have done simulation experiment in laboratory cloud chamber and have observed condensation and water drops formation. They also succeeded in producing tiny water particles in moderately humid air in an altitude of 45 to 75m of the atmosphere by terawatt mobile laser. But the droplets were about a hundred times too small to fall as raindrop; instead, they remained suspended in the air. The team feels that it is feasible to get larger droplets if the power of laser is increased to petawatt (10^{15}watts) or exawatt(10^{18}watts). ***They further say that the effectiveness of this method is much easier to gauge than traditional cloud-seeding techniques and that it could provide to be a practical means of triggering rainfall****". (Search Google as 'Laser makes rain, heavily'2010).* A group of Scientists from Florida University also observed water drops formation by high power laser shooting experiment. It appears from the above ***that laser has not yet succeeded in producing***

artificial rain. In this project proposal, a novel method is described to create artificial rain by laser systems.

10.2: Condensation Is the Basic Need for Water Drops Formation:

IRRA Scientist Group has been successful experiments in laboratory, for 'condensation is the basic need for water drop formation'. That condensation is the basic need for water drop formation can be understood by taking two glasses, one filled with normal water and another with ice pieces. After sometime one can observe water droplets on the outer surface of the glass which contains ice but not in the other. This is due to the condensation process that occurs around the ice glass. So, IRRA Scientist Group, proposed a laser system for this research project, to create artificial lightning by initiation of endothermic reactions, similar to natural lightning phenomena, for artificial rainmaking. As a result of these reactions, temperature falls, condensation takes place, seeding occurs and it rains in an analogous way similar to rains created by lightning. This process has been practically proved in the laboratory and atmosphere as "production of ozone and nitrogen oxides by laser filamentation". A number of scientific and practical tests have been conducted in cloud chambers and in the atmosphere to prove this hypothesis, as "Laser induce condensation and water drops formation by Laser shooting in the laboratory cloud chambers as well as in the atmosphere".

Now the question arises that what are the conditions required for condensation. This means that, only and only, endothermic reactions are responsible for condensations, which also produces NO (Nitrogen Oxides) and O_3 (Ozone) after shooting laser beams triggering endothermic reactions, condensation and precipitation. These tiny water drops act as a natural seeding process, due to acceleration and tribulation by wind force in the atmosphere, to form another set of raindrops with heavy rainfall as lightning rain. Endothermic reactions are responsible for condensation. Condensation is the basic need for water drop formation as above laboratory experiment, "Condensation is the Basic Need for Water Drops Formation".

10.3: Scientific, & Practical Background on Rainmaking Technology:

This novel Rainmaking Technology can be used for white, warm clouds too which get converted into black rainy clouds for rain enhancement. As well as, water drops formation by high power Laser shooting, "Laser makes rain".

The IRRA scientist Group has already demonstrated "Innovative Rainmaking Technology" in the Laboratory cloud's chamber successfully, and results have been published in "Indian Journal of Science & Technology", http://indjst.org, vol.1, No.6 (2008).

Now, the project proposal, the design of the laser system, the Budget estimate, and the work plan are ready with IRRA Scientist Group, India. IRRA Scientist is ready for demonstration and collaboration with the Government for funding purposes for project proposal on "Artificial rainmaking by Peta Watt double Laser system from ground initiation of Endothermic Reactions, as similar natural lightning phenomena in the atmosphere"

In this experiment, Peta Watt (10^{15}watt) double laser creates artificial lightning in the atmospheric cloud's regions from ground. These white warm clouds are converted into black rainy clouds with natural seeding for rain enhancement in the atmosphere.

A laser pulse will be sent to the cloud to initiate endothermic reactions which will create lightning phenomena, as in nature, mentioned above. For example, a German-French group has used a femtosecond–terawatt laser to obtain "Laser-assisted water condensations in the atmosphere". They have succeeded in obtaining raindrops from an altitude of 45m to 75m of the atmosphere. In this experiment, intensity of Laser system can't reach above 75m an altitude in the atmosphere. IRRA Scientist propose "Artificial rainmaking by Peta Watt (10^{15}watt) double Laser system from ground initiation of Endothermic Reactions, as similar natural lightning phenomena in the atmosphere"

*This novel Rainmaking Technology can be used for white, warm clouds too which get converted into black rainy clouds for rain enhancement. *As well as, water drops formation by high power Laser shooting.*

"Laser makes rain" by Florida University, Scientist experimentally observed.

**As per report, a group of European scientists working on artificial rain said in 2010, "Firing extremely powerful laser pulses through humid air can stimulate the formation of clouds, according to a team of European scientists. They say that the effectiveness of this method is much easier to gauge than traditional cloud-seeding techniques and that it could provide to be a practical means of triggering rainfall". (Search Google as 'Laser makes rain, heavily'2010). In Indian ancient tradition, there is a mention in Vedshastra that "Fire arrows are sent towards the atmospheric clouds which is responsible for immediate rainfall"*

Our system could be Peta Watt (10^{15}watt) double Laser system from ground, its fundamental wavelength could be ~800nm. The pulse will have energy of ~500mJ, 120fs and repetition frequency of 10Hz. The laser pulse has to propagate with almost high peak intensity in atmospheric clouds. It works when more than 65% humidity is present in the atmosphere. Our findings could be used by scientists and engineers to create artificial rain through a new method. The results could be of immense benefit to human being.

10.4: How Will Laser Create Artificial Rain? Theory:

For creation of rain, according to well established meteorology theory (can be found in any textbook of meteorology), steps are the following:

First (i) creation of low temperature →then (ii) condensation then (iii) seed (CCN) formation →then (iv) tiny water drops formation and rain occur in the atmosphere.

The present methodology is to send laser pulse to the cloud region of the atmosphere to create high temperature. This high temperature will break the bonds of O_2 (21%) and $N_{2\,(74\%)}$ as follows:

$N_2 : N \equiv N \rightarrow N^* + N$ ---- (1)

$O_2 : O = O \rightarrow O^* + O$ ----- (2)

In this processing N and O in excited state (N^*, O^*) will be created. These excited N^* and O^* are very unstable and immediately come to the ground state through following reactions.

$N^* + O_2 \rightarrow \quad NO + O\ \Delta H\ (43.2kcal/mol)$ …… …….. (3)

$O^* + O_2 + M \rightarrow O_3 + M\ \Delta H\ (67.7kcal/mol)$ …… (4)

The occurrence of reactions 3 and 4 and formation of NO and O_3 have confirmation from NASA laboratory experiments [Sanders et al. 2003]. Formation of O_3 and NO, after laser shot, has been observed in laser experiment in atmosphere (Petit et al. 2010). The reactions 3 and 4 are endothermic and therefore, they need heat energy (amount mentioned in brackets) which is absorbed from the cloud region. As a result, temperature of the cloud region falls (first step of rain formation is achieved and then other steps follow), condensation takes place, seeds (CCN) will be formed, and tiny water drops will be created. These tiny water drops may act as natural seeds to form another sets of rain drops. This chain process will result in rainfall. Ozone and Nitric oxide, O_3 and NO (formed in reactions 3 and 4) will undergo further reaction to form HNO_3 particles other nitrogen compounds, which will bind water molecules together to create water droplets. These water droplets again will act as natural seeds to form another sets of rain

drops. In the atmosphere, due to turbulence, small water drops coalesce and form big rain drops. In addition, ions N_2^+ and O_2^+ and electrons formed by cosmic rays can create complex

hydrated heavy positive and negative ions... $HNO_3^-.(H_2O)_n$ (where the value of n could be as large as 50) which can also act as seed to create rain.

In short, to create artificial rain by laser, endothermic reactions are to be generated in the cloud region. Its has-been shown earlier how much heat energy is absorbed by endothermic reactions from atmospheric clouds (Chopkar 1993a, b; Chopkar and Chakrabarty 2008; Chakrabarty et al. 2010; Chopkar et al. 2010). The energy required to break bonds of 1 molecule of N_2 and 1 molecule of O_2, $= 2.25x10^{-18}$ Joule. A laser pulse of energy 500mJ can dissociate a column of N_2 and O_2 containing

($\sim$0.5/2.25^{-18}) $\sim$ 10^{17}molecules which is much higher than the density in the atmosphere.

Laser can be operated from the ground as well as from an aircraft. In former case, laser pulse has to propagate to a height of $\sim$1km (cloud height) from the ground. There will be attenuation of energy in this propagation. Kasparian et al. (2012) experimented with terawatt laser from the ground and observed tiny raindrops in an altitude of 45m to 75m of the atmosphere. To create large water droplets at higher altitudes, the group feels that laser power has to be peta-watt (10^{15}watt) or hexa-watt (10^{18}watt). If laser is operated from aircraft, then attenuation of energy will be less. In that case, laser power can reach the cloud region without much attenuation. It can also cover large area and can move to any place. Turbulence created by the aircraft in the atmosphere can also create small water drops which would collide with each other and form big rain drops.

In this research paper, IRRA Scientist propose "Artificial rainmaking by Peta Watt (10^{15}Watt) double core Laser system from ground initiation of Endothermic Reactions, as similar natural lightning phenomena in the atmosphere"

Artificial rain making methods help to increase the green life and oxygen and decreasing the pollution. Thus, these methods play major role in reducing drought and increasing the quantity of drinking water in future.

10.5: Methodology for Artificial Rainmaking by Laser System from Ground:

In this experiment, Peta Watt (10^{15}watt) double core Laser system creates artificial lighting in the atmosphere up to 1.2 Km to 2.7 Km altitude in the white warm cloud's regions. These white warm clouds are converted into black rainy clouds with natural seeding for rain enhancement in the atmosphere.

A laser pulse will be sent to the cloud to initiate endothermic reactions which will create lightning phenomena, as in nature, mentioned above. The laser technology for this purpose, though not fully developed, yet exists. For example, a German-French group has used a femtosecond–

terawatt laser to obtain "Laser-assisted water condensations in the atmosphere". They have succeeded in obtaining raindrops from an altitude of 45m to 75m of the atmosphere.

*This novel Rainmaking Technology can be used for white, warm clouds too which get converted into black rainy clouds for rain enhancement. *As well as, water drops formation by high power Laser shooting. "Laser makes rain" by Florida University, Scientist experimentally observed.*

**As per report, a group of European scientists working on artificial rain said in 2010, "Firing extremely powerful laser pulses through humid air can stimulate the formation of clouds, according to a team of European scientists. They say that the effectiveness of this method is much easier to gauge than traditional cloud-seeding techniques and that it could provide to be a practical means of triggering rainfall". (Search Google as 'Laser makes rain, heavily'2010). In Indian ancient tradition, there is a mention in Vedshastra that "Fire arrows are sent towards the atmospheric clouds which is responsible for prompt rainfall"*

Our system could be Peta Watt (10^{15}watt) double core Laser system, its fundamental wavelength could be ~800nm. The pulse will have energy of ~500mJ, 120fs and repetition frequency of 10Hz. The laser pulse has to propagate with almost high peak intensity over a distance of ~1km. It works when more than 65% humidity is present in the atmosphere. Our findings could be used by scientists and engineers to create artificial rain through a new method. The results could be of immense benefit to human being.

This laser system can be operated from ground as well as from an aircraft. In this experiment, laser system can be operated from ground, Innovative rainmaking technology in the atmosphere as shown in Fig.No.1, in this demonstration, Peta Watt (10^{15}watt) pulse laser with double coarse Laser (1) primary laser and (2) secondary Laser as shown in fig.No.1.

A laser pulse has to be sent in the atmosphere to reach white warm clouds regions about 1.2 Km to 2.7 Km altitude. It will initiate endothermic reactions which will create lightning phenomena and rain as in nature. These white warm clouds will be converted in to black rainy clouds with natural seeding for rain enhancement in the atmosphere.

This laser system can be operated from ground as well as from aircraft. Plans to operate from ground are shown in figures No.1. When operated from ground, covered area on the ground could be ~5 Km to 7 Km in radius. Fig.1 shows a plan to use two lasers. In this figure, Peta Watt (10^{15}watt) double Laser pulse laser with double coarse laser (1) primary laser and (2) secondary laser is shown. In this system, secondary laser energy is used for laser pulse travelling purpose and primary laser energy is used for creating artificial lightning in the upper atmosphere up to 1.2 Km to 2.7Km altitude for initiation of endothermic reactions, lot of heat energy absorbs from surrounding atmospheric clouds, condensations take place, water drops formations in the atmosphere. It's used as natural seeding process, to form another sets of raindrops, chains process occurs for rain enhancement in the atmosphere.

Innovative rainmaking technology by laser system from ground level can be used in the field of farmers. When the crops are on the verge of drying due to lack of rain since one to two months, this experiment can come to an aid. It can cover an area of more than 7 km of radius in a wind direction in the atmosphere as per Fig.No.1. In this experiment, two cylinders for Laser sending and travelling, one is vertical 750m height, another inclined cylinder 500m with rotating around in 360 angles horizontally as shown Fig.No.1.

For this experiment, hilly area must be selected on the field of farmer, with high power electric supply point already managed near the experiment site. High power laser system must be placed in a 12-wheel truck so that it becomes easy to transport it from one place to another. When the upper atmosphere is cloudy and more than 65% humidity is present in the atmosphere, the experiment can be started in the field of the farmer. We will measure atmospheric parameters such as humidity, temperature, pressure, wind velocity, wind direction, etc. on ground level as well as in the upper level. For this purpose, a separate unit/department must be established as "Measuring & Maintains Department". After the success of experiment, all related data, must be put it in software computer, for analysis and conclusion, with fixed perfect Laser design for maximum rainmaking / rainfall in the atmosphere.

In this way, we will use Laser system from ground as Peta Watt (10^{15}watt) double Laser power with double Laser means primary and secondary laser). Secondary Laser power will used for travelling purpose and primary Laser power used for creating artificial lightning in the atmosphere for rainmaking. Similarly, Laser power has to be peta-watt (10^{15}watt) or hexa-watt (10^{18}watt) should be used for creating artificial lightning in the atmosphere for rainmaking from ground level which is send as Laser pulses length 2.5m to 5.0m in repetition process in upper cloud regions as shown in Fig.No.1.

Demonstration on project proposal of Innovative Rainmaking Technology by Laser system from Ground level for Rain enhancement in the Atmosphere by IRRA Scientist Group.

"Artificial Rainmaking Methodology by Laser system from Ground", it's most useful for green revolution in the whole world for all human beings.

In this way "Innovative Rainmaking Technology by Laser system from Ground", can be used for green revolution in the whole world for all human beings.

10.6: Discussion:

Innovative Rainmaking Technology is scientifically and practically proven as "Laser induces condensation and water drops formation in Laboratory cloud's chamber as well as in the atmosphere.

However, according to Kasparian group, a laser pulse shot in the atmosphere ionizes N_2 and O_2

$N_2 + h\nu \rightarrow N_2^+ + e^-$ ………..... (7)

$O_2 + h\nu \rightarrow O_2^+ + e^-$ …....... (8)

They have observed lightning phenomenon in the laboratory cloud chamber as "Laser induced condensation and water drops formation in the laboratory cloud chamber by Femtosecond –Terawatt mobile laser system". Kasparian (2012) group says that it is the ionized species N_2^+ and O_2^+ which produce rain. But these two species are of micro size which *cannot act as seeding agents. Also, N_2^+ and O_2^+ radicals are not observed by Kasparian group in laser filamentation experiments but production of O_3 and NO has been observed by them in laser filamentation experiments. Experiments of Kasparian group finds condensation and water drop formation and they say in their ionization theory that N_2^+ and O_2^+ act as seeding agent. They also say "Mechanism of laser–induced condensation involves photo dissociation, in which photons break down atmospheric compounds in the atmosphere". This process produces Ozone and Nitrogen oxides, which lead to the formation of Nitric acid particles that bind water molecules together to create water droplets." But there is no seeding and condensation and water drop formation is not due to seeding. Small water drops formed by laser in the laboratory cloud chamber are due only to endothermic reactions (cooling) and this is obvious. In the atmosphere, due to acceleration and turbulence, these small sized water drops coalesce to form big rain drops. These rain drops act as a natural seeding process to form different sets of rain drops; this chain process continues with heavy rainfall.*

Calculation for the energy required for dissociation is almost half of that required for ionization
The energy of a laser beam of wavelength λ is hν (ν = 1/λ and h is Planck's constant). We will shoot laser pulse in the atmosphere and dissociate (break bonds of) N_2 and O_2 as follows:

$N_2 + h\nu \rightarrow N^* + N$ …… (1)

$O_2 + h\nu \rightarrow O^* + O$ …… (2)

Bond energy of N_2 = 226 kcal/mole.

1 cal = 4.184 Joule, Avogadro number = $6x10^{23}$

Therefore, energy required to break 1 molecule of

$N_2 = 226x10^3x4.184/ (6x10^{23}) = 1.58x10^{-18}$ *Jou*

Bond energy of O_2 = 96 kcal/mole.

Therefore, energy required to break 1 molecule of

$O_2 = 96x10^3x4.184/ (6x10^{23}) = 0.67x10^{-18}$ *Joul*

So, the total energy required for breaking 1 molecule of N_2 *and 1 molecule of* O_2 *will be* $(1.58x10^{-18} + 0.67x10^{-18})$ = $\underline{2.25x10^{-18}}$ *Joule.*

When a laser pulse is shot in the atmosphere, it may ionize N_2 *and* O_2 *as follows:*

$N2 + hv \rightarrow N_2^+ + e$*........ (3)*

$O2 + hv \rightarrow O_2^+ + e$*........(4)*

Ionizing potential of N_2 = 15.58 ev = $2.49x10\text{-}^{18}$*Joule*

Ionizing potential of O_2 = 12.2 ev = $1.95x10^{-18}$*Joule*

So the total energy required ionizing 1 molecule of N_2 *and 1 molecule of* O_2 *is* $2.49x10^{-18}$ *Joule +*$1.95x10^{-18}$*Joule* = $\underline{4.44x10^{-18}}$ *Joule.*

The above calculation shows that the energy required dissociating 1 molecule of N_2 *and 1molecule of* O_2 *is about half of that required to ionize them.*

The energy required for dissociation is almost half of that required for ionization.

That means energy is first used up for dissociation, then the remaining energy (which may not be sufficient for ionization of N_2 *and* O_2*) is delivered for ionization of* N_2 *and* O_2*. Hence dissociation takes place and not ionization. The Kasparian group does not talk about*

dissociation. It is not only near the IR laser system, Yoshihara, et al. (2007), have discussed in their paper the possibility of creating artificial rain by using UV lasers. Our methodology is to send laser pulses to cloud regions to break bonds of O_2 *and* N_2 *(by reactions 1 and 2), create endothermic reactions and condensation (by reactions 3 and 4) and produce rain in the similar way as in lightning. There is attenuation of energy in operating the laser through Aircraft.*

Kasparian's group suggests an increase of laser power to petawatt (10^{15}*watt) or exawatt (*10^{18} *watt) to create large water droplets. We will operate from an aircraft in the same way as spraying chemicals from Aircraft. A laser pulse of energy 500mJ is capable of dissociating a* *column of N_2 and O_2 containing $(\sim 0.5/2.25^{-18}) \sim 10^{17}$ molecules which is much higher than the density in the atmosphere.*

10.7: Acknowledgement:

We express our sincere thanks to scientists: Prof Mamata R. Lanjewar (RTM University Nagpur), Prof A. P. Deshpande (Principal, Science College), Prof. Padmanabhan Murthy (J.N. University, New Delhi), Prof Umesh Kulshetra (J.N. University, New Delhi), Prof K. S. Korgaokar (Pune University, Pune), Dr. A.L. Agarwal (NEERI Nagpur), Dr. Nitin Saraf (B.D. Engineering College, Sewagram), Dr. K. M. Kharate, (GM Engineering College & Research Centre, Sheogao), Dr. Mrs. Sumanlatha Pandey, Mr. Shyam Ujjaninkar (Engineer), Mr. Bhola Katare, Bhandewadi, (Jay Shri Jaganatha Namo Namha), Mr.Vithal Wagh, Waigegao,Sadguru Parmhans shri Ramchandra Maharaj Namo Nemaha. Mr. Rajesh Iyengar (P. N. College, Nanded). Dr. K.R. Gangakhedkar, (P. N. College, Nanded). Ex. Prof. Ratnakar Lanjewar (HOD, Chemistry Department), Science college, Nagpur. Nanded) for their valuable suggestions and guidance for project proposal.

10.8: Additional Uses:

This method can be used for rain harvesting by "rain drain". When huge clouds are present above a lake or dam, a laser beam can be shot into the cloud region; then with blast of clouds heavy rainfall will occur to fill the lake or dam for future use of water. This method can also be used to reduce pollution of the atmosphere by spraying artificial rain on the polluted city. Another use of this method is to stop excess rainfall. Low intensity laser pulse shot into the cloud region will evaporate the clouds from the excess rainfall area. This method can also be used to drive away the rain cloud from the region where rain is not needed.

10.9: Conclusion:

"IRRA Scientist Group propose Peta Watt (10^{15}Watt) double core laser system from ground of specification: 10^{15}watt, 800nm, 500mJ, 120fs and 10Hz for this research project". The results could be of immense benefit to human being.

It is shown in this project proposal that by initiating endothermic reactions in the cloud region of the atmosphere by a laser, artificial rain can be created. Laser may have the following specification: 10^{15}watt, 800nm, 500mJ, 120fs and 10Hz for operation from ground level. This method is economical (one time investment), harmless, eco-friendly and can be switched on and off when desired. It can be used at any place and at any time. It can also be used to fill lake or dam for storing rain water for future use (rain harvesting), to reduce pollution by spraying artificial rain on the polluted city, to stop rain in the region where it is not wanted or where rainfall is in abundance. It may not be out of place to state here that in the holy Hindu book "Mahabharata", there is a mention that by firing arrows in the atmosphere, rain was created by Arjun to quench the thirst of God "Bhishma".

Artificial rain making methods help to increase the green life and oxygen and decreasing the pollution. Thus, these methods play major role in reducing drought and increasing the quantity of drinking water in future.

10.10: References:

Braun, A., Korn, G., Liu, X., Du, D., Squier, J., and Mourou, G., 1995. "Self-channeling of high- peak power femtosecond laser pulses in air." Opt. Lett., vol. 20, pp. 73-75.

Carls, J. C. and Brock, J. R., 1987. "Explosion of a water droplet by pulsed laser heating." Aerosol Sci. Technol., vol. 7, pp. 79-90.

Chopkar, S. K. and Chakrabarty, D. K., 2008. "Artificial rainmaking system in a way of natural phenomena." Indian J. Sci. Technol., vol. 1, pp. 1-5. Available: http://www.indjst.org

Chakrabarty, D. K, S. K. Chopkar and N. N. Purkait, "Femtosecond terawatt laser system to produce artificial rain" presented at 4th international conference on computer and devices for communication, CODEC 2009 at Kolkata in December 2009 and published in IEEE X-plore digital library, INSPEC Accession No. 11136783, Print ISBN 978-1-4244507329, February 2010.

Chopkar, S. K., "Effect of endothermic reactions associated with lightning on atmospheric chemistry" *Indian J. Radio Space Phys.22*, 128-131, 1993a.

Chopkar, S. K., *American Meteorological Society in Meteorological & Geo-astrophysical Abstracts,44*, October 1993 No.10 (44.10-555) 1993b.

Chopkar, S. K., D. K. Chakrabarty, S. M. Sonbawane, R. L. Bakal, P. S. Chopkar, P. Hariom and J. S. Pimplkhute, *International J. of Meteorology, 35(355)*, 363-370, 2010.

Golde, R. H., *Lightning, vol. 1, Physics of Lightning*, Academic Press, London, 1977.

Kasparian, J., R. Sauerbrey and S. L. Chin, The critical laser intensity of self-guided light filaments in air, *Appl. Phys. B 71*, 877-879, 2000.

Kasparian, J., M. Rodriguez, G. Mejean, J. Yu, E. Salmon, H. Wille, R. Bourayou, S. Frey, Y. –B. Andre, A. Mysyrowicz, R. Sauerbrey, J. –P. Wolf and L. Woste, White-light filaments for atmospheric analysis, *Science, 301*, 61-64, 2003.

Kasparian, J, P. Rohwetter, L. Wöste and J. –P. Wolf, Laser-assisted water condensation in the atmosphere: a step towards modulating precipitation, *Journal of Physics D: Applied Physics,45 (29),* 2012.

Mejean G., Ackermann R., Kasparian J., Salmon E., Yu J., Wolf J. -P., Rethmeier K., Kalkner W., Rohwetter P., Stelmaszczyk K. and Woste L., (2006) Improved laser triggering and guiding of megavolt discharges with dual fs-ns pulses, App. Phys. Letts.,88, 021101

Petit, Y., S. Henin, J. Kasparian and J. –P Wolf, Production of ozone and nitrogen oxides by laser filamentation, *Appl. Phys. Lett. 97*, 021108, 2010.

Rohwetter, P., J. Kasparian, K. Stelmaszczyk. Z. Hao, S. Henin, N., Lascoux, W. M. Nakaema, Y. Petit, M. Queisser, R. Salame, E. Salmon, L. Woste and J. –P. Wolf, Laser-induced water condensation in air, doi: 10.1038/nphoton.2010.115, 2010.

Sander, S. P., R. R. Friedl, D. M. Golden, M. J. Kurylo, R. E. Huie, V. L. Orkin, G. K. Moortgat, A. R. Ravishankara, C. E. Kolb, M. J. Molina and B. J. Finlayson-Pitts, Chemical kinetics and photo-chemical data for use in atmospheric studies, *NASA JPL publication,* pp. 02-25,

Yoshihara, K., Y. Takatori, K. Miyazaki and Y. Kajit, Ultraviolet light-induced water- droplet formation from wet ambient air, *Proc. Jpn. Acad. Sci. B 83*, 320.

By

Innovative Rainmaking Research Association (IRRA Scientist Group)

Sevagram- 442102, Dhanvantri Nager-30, Warud, Dist.Wardha Maharashtra, India, E-skc.arr@rediffmail.com, E-irra.scientistgroup@gmail.com, Website-www.irraindia.org

10.11: Figure:

Demonstration on Rainmaking Technology by Peta watt double Laser system from Ground:

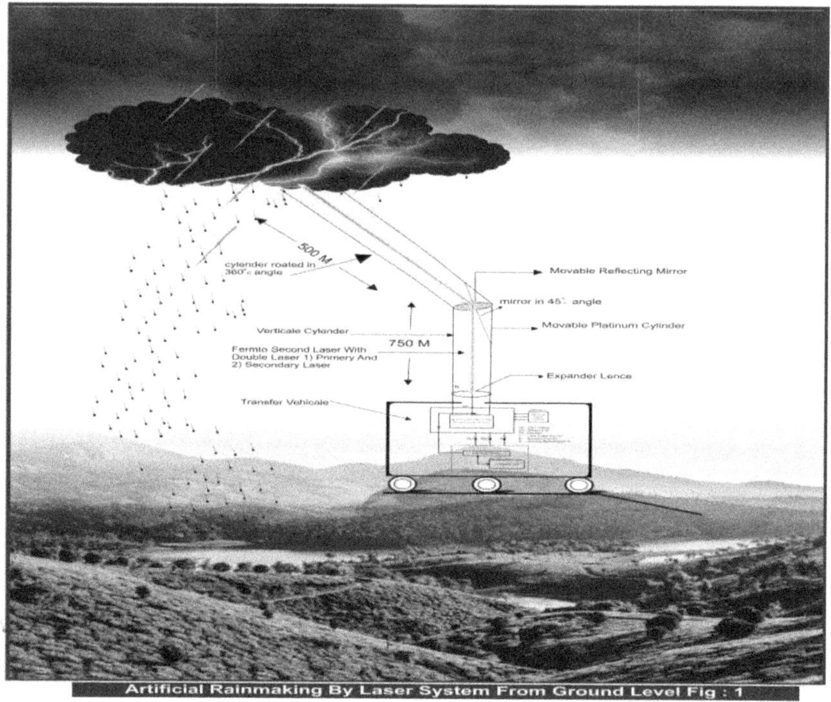

Artificial Rainmaking By Laser System From Ground Level Fig : 1

Fig. 1

10,12: Budget for Three Years:

"Artificial rainmaking by Peta Watt double Laser system from ground initiation of Endothermic Reactions, as similar natural lightning phenomena in the atmosphere"

(Grand Total Estimate for Rainmaking Project proposal (A+B+C+D+E) – 9.02 Cr in Indian Rs/ USA 0.72 million Dollar)

Demonstrated/experiment for Artificial rainmaking by Peta Watt double Laser system from ground for three years, project proposal estimates cost 9.02 Cr./USA dollar 0.72 million.

Sr.	Item	Budget of Rain Making Project			
		1st Year	2nd Year	3rd Year	Total (In Lakh)
A	***Budget for Salaries of Staff:***	**54**	**54**	**54**	**01,62**
B	***Budget of Equipment Purchase with maintains etc.***	0390	00.05	00,05	04,00
C	***Budget for Consumable Item:***	75	70	70	**02,15**
D	***Budget for Travel Expenses:***	***00,40***	***00,25***	***00,25***	**00,90**
E	***Budget for Patent Filling Expenses***	***00,15***	00,10	00,10	**00,35**
	Total	**574**	**164**	**164**	**09,02**

(Grand Total Estimate for Rainmaking Project proposal (A+B+C+D+E) – 9.02 Cr in Indian Rs/ USA 0.72 million Dollar)

Demonstrated/experiment for Artificial rainmaking by Peta Watt double Laser system from ground for three years, project proposal estimates cost 9.02 Cr./USA dollar 0.72 million.

A) Budget for Salaries of Staff:

SN	Designation	Pay Scale	Monthly	Nos Of Staff	Budget for Salaries of Staff			
					1st Year	2nd year	3rd Year	Total
1	**Senior Scientist (Project Manager)**	**Class 1**	1,20,000	1	14,40,000	14,40,000	14,40,000	**43,20,000**
2	**Assistant Scientist**	**Class 2**	80,000	2	19,20,000	19,20,000	19,20,000	**57,60,000**
3	**Junior Scientists**	**Class 2**	40,000	2	09,60,000	09,60,000	09,60,000	**28,80,000**
4	**Lab Helper with Skilled Labor**	**Class 3**	20,000	3	07,20,000	07,20,000	07,20,000	**21,60,000**

5	Non- Skilled Labor	Class 4	10,000	3	03,60,000	03,60,000	03,60,000	10,80,000
	Total				54,00,000	54,00,000	54,00,000	01,62,00,000

B) Budget of Equipment Purchase:

SN	Particular	Model	Make	Estimate Cost (In Lakhs)
1	High power laser Instrument with Transportation (Femto-second high power Laser Instrument)	2 Numbers with multiple laser release instruments	Imported	1,60
2	Construction cost for Camp/ guest house /office temporary Shed		Indigenous	70
3	High power electric supply point /Fuel Materials		Indigenous	20
4	Precipitation & Atmospheric parameters measuring instrument etc.		Indigenous	40
5	Purchase Mobile Van/ official materials etc.		Imported	60
6	Scientific equipment purchases as per requirement		Imported	40
	Total			390

C) Budget for Consumable Item:

Budget of Consumable Item					
Sr.	**Consumable Items**	**1st Year**	**2nd Year**	**3rd Year**	**Total (Lakhs)**
1	**Electric supply for high power laser/Fuel etc.**	10	05	05	**20**
3	**Stationary & boarding, mese expenses, another misalliance**	45	45	45	**135**
4	**Purchase Equipment for project, office etc.**	10	10	10	**30**
5	**Chemicals for testing or Analysis**	10	10	10	**30**
	Total	**75**	**70**	**70**	**215**

D) Budget for Travel Expenses:

Budget of Travel Expenses					
SN	**Description**	**1st Year**	**2nd Year**	**3rd Year**	**Total (Lakhs)**
1	*Conference for advance technology for discussion and suggestion*	*20*	*10*	*10*	*40*
2	*Visit of top scientists to project for various Suggestion*	*10*	*10*	*10*	*30*
3	*Visit for other institute for research study &discussion*	*10*	*05*	*05*	*20*
	Total	*40*	*25*	*25*	*90*

E) Budget for Patent Filling Expenses:

Budget of Patent Filling Expenses & other					
SN	**Description**	**1st Year**	**2nd Year**	**3rd Year**	**Total (Lakhs)**
1	*National & International Patent file expenditure and advertisement etc.*	*05*	*05*	*05*	*15*
2	*Consultant fees / Expert guidance Service charges*	*10*	*05*	*05*	*20*
	Total	*15*	*10*	*10*	*35*

10.13: Practical Result Shows It's Positive To "Innovative Rainmaking Technology" Scientifically & Practically Proven As ...

1. ***"Laser induces condensation and water drops formation in the cloud's chamber" By J. Kasparian and*** A team, called terra-mobile-group (TMG), consisting of scientists from Switzerland, Germany and France, have been trying to create artificial rain by laser ***(results published in international journal……)***

2. ***"Laser induce condensation and water drops formation in the atmosphere up to 75m altitudes":*** *This process has been practically proved in the laboratory as* ***"Production of ozone and nitrogen oxides by laser filamentation".*** *It is believed that "Laser photons photo-dissociate atmospheric compounds N_2 and O_2 form ozone (O_3) and nitrogen molecules (NO).* ***"Increase of O_3 and NO concentration after lightning has also been experimentally observed".*** *This lightning phenomenon created through artificial lightning by a plasma laser pulse or laser system can produce rain in the atmosphere has been practically proved as* ***"Laser-induced water condensation in the air".*** *Scientists have succeeded in obtaining raindrops from an altitude of 75m of the atmosphere by terawatt mobile laser.*

We well know, NO and O_3 are endothermic reactions, lot of heat energy is required for these reactions which is subtracted from surrounding

clouds…..condensations take place…..condensation is the basic need for water rain drops formations (IRRA Scientists practically proved in Lab).These rain drops act as natural seeding process by wind forces due to acceleration & turbulences to form another sets of rain drops for enhancing rainfall.

3. ***"Laser makes rain" by Florida University Scientists practically observes …. Ref. (Google search). They further say that the effectiveness of this method is much easier to gauge than traditional cloud-seeding techniques and that it could provide to be a practical means of triggering rainfall"***. By European Scientists (*Search Google as 'Laser makes rain, heavily'2010)*

4. ***Several attempts have been made by various researchers to create artificial rain by laser. Golde (1977) from a number of radar observations has reported that intense precipitation is not even present in the clouds before the first discharge but develops abruptly in the same region after release from which the lightning flashes originate.***

"Innovative Rainmaking Technology" is a natural phenomenon similar to natural lightning phenomena……

5. ***"Artificial Rainmaking System in a way natural phenomenon" research paper published in journal search on http://www.indist.org,vol.1No.6(Nov. 2008), In this research paper, IRRA Scientists has been published the practical result in a laboratory chamber ….***

6. ***In Indian philosophy, there is mention in 84 Sukkot "Agni ban Badal me Chhodake , parjany vurishti kihi" says Veda Shastra. Veda Shastra's thoughts are true facts in Adhayatm. Means ….as "Fire Rocket launch in the atmosphere for rainfall" …..***

7. ***Now, IRRA Scientists would like to demonstrate "Innovative Rainmaking Technology" on a large scale in the atmosphere, for that's ….."IRRA Scientist Group propose laser system of specification: 10^{15}watt, 800nm, 500mJ, 120fs and 10Hz for this research project".*** *The results could be of immense benefit to human beings.*

10.13: Work-Plan

Demonstration on Rainmaking Technology project proposal for Three Years Work Planning

First Year		
1-4 Months	***5-8 Months***	***9-12 Months***
Publishing the advertisement for the appointment of scientists, skilled and non-skilled workers in the reputed newspaper.	***Inviting experts from the concerned area for discussion on the project***	***Organization of two days national conference on the theme of the project***
Demand for different instruments to different venders	***Comparative statements of received quotations for the different instruments.***	***Accessories, assembly and instalment of received instruments on the mobile van and their testing***
Selection of proper site and arrangement of electricity, road, guest house etc.	***Purchase order of proper instrument to proper manufacturing company***	
Second Year		
1-4 Months	***5-8 Months***	***9-12 Months***
carrying experiments on the selected site and recording the observations	***Inviting experts from the concerned areas for table discussion on the observations taken***	***Analysis of the observations taken***
	carrying experiments on the selected site and recording the observations as per the direction given by the experts	***Interpretation of the data***
		Publication of the result in the reputed national and international journals

Third Year		
1-4 Months	***5-8 Months***	***9-12 Months***
Analysis of the observations taken	***Compilation of the data in computer software and commercialization of the same for the benefit of society.***	***Filing the patent for the artificial rainmaking system***
Interpretation of the data		***Appeal to the Indian government and other countries to utilize this project for the benefit of Humankind.***
Publication of the result in the reputed national and international journals		

10.0 "Two Project Proposal on Rainmaking Technology for Demonstration from Ground & onboard Aircraft in the atmosphere".

(10.2 Project proposal For Rainmaking from Aircraft)

"Artificial rainmaking by Tera Watt (10^{12} Watt) mobile Laser system from Aircraft, initiation of Endothermic Reactions, as similar natural lightning phenomena in the atmosphere"

S. K. Chopkar, D. K. Chakrabarty, K. R. Gangakhedkar, Umesh Kulshetra., Prof. Dr. Jzsef Dr. Steier

Sewagram- 442102, Dhanvantari Nager-30, Wardha, Maharashtra, India.

E-skc.arr@rediffmail.com, E- irra.scientistgroup@gmail.com, Website-www.irraindia.org

Abstract:

In the atmosphere, after lightning, precipitation is formed and heavy rainfall occurs. This is a well-known process. This natural lighting phenomenon has been practically demonstrated in the laboratory cloud chamber, as a "Laser-induced condensation & water drop formation" and "Water drops formation by each and every laser shot in the cloud chamber". In this process, lightning/laser creates high temperature which breaks the bonds of N2 and O2 to form excited N and excited O*. The total heat energy of lightning/laser is completely utilized for breaking the bonds of N2 and O2 (Chopkar, 1993). This excited N* and excited O* moves to the new place by wind and undergoes reactions to form NO and O3 which are endothermic reactions. The heat energy required for these reactions is taken from the surrounding atmospheric clouds.*

As a result of these reactions, the temperature falls, condensation takes place, and raindrops form, these raindrops act as a natural seeding process, to form another set of raindrops, seeds are created and rain occurs in an analogous way similar to rains created in nature by lightning.

In this process, white clouds convert into black rainy clouds for rainmaking in the atmosphere. This process has been practically proven in the laboratory as "Production of ozone and nitrogen oxides by laser filamentation". It is believed that "Laser photons photo-dissociate atmospheric compounds N2 and O2 form ozone (O3) and nitrogen molecules (NO). "An increase in O3 and NO concentration after lightning has also been experimentally observed".

This lightning phenomenon created through artificial lightning by a plasma laser pulse or laser system can produce rain in the atmosphere and has been practically proved as "Laser-induced water condensation in the air".

Scientists have obtained raindrops from an altitude of 75m of the atmosphere by terawatt mobile laser. "IRRA Scientist Group propose laser system of specification: 10^{12}watt, 800nm, 500mJ, 120fs and 10Hz for this research project". The results could be of immense benefit to human beings.

Keywords: Artificial rain; Atmospheric cloud; Condensation; Endothermic reactions;

Laser system; Precipitation; Raindrops; Rainfall; Natural Seeding.

10.0 Introduction:

Rain plays a great role in national economy by influencing the agriculture yield. Nevertheless, rain is a natural phenomenon and it does not fall as and when man needs it and hence, man has been trying to create artificial rain for the past many years. At present, the process, which is used for creating artificial rain, is seeding. In this process, chemicals such as silver iodide, calcium chloride or sodium chloride are used as seeds. They are spread from the aircraft in the cloud region. Nucleation starts on these chemicals, which lead to the precipitation and then rain. This process has been tried in South Africa,

Thailand, Japan, Mexico, Brazil and some parts of India. However, this process, on most of the occasion fails. Besides, it is harmful to humankind because it brings harmful

chemicals on earth along with the rain. It is, also, expensive. In this paper, we propose a laser system to produce artificial rain. Our system has advantages over the seeding process that it does not pollute the environment; it is a onetime investment and is less expensive. In addition, it can be targeted to even warm white clouds, which are not rainy (in seeding process only black rainy clouds are targeted).

10.1 Practical Evidences:

Several attempts have been made by various researchers to create artificial rain by laser. Golde (1977) from a number of radar observations has reported that intense precipitation is not even present in the clouds before the first discharge but develops abruptly in the same region after discharge from which the lightning flashes originate.

Carls and Brock (1987) heated the atmosphere by a laser pulse up to 1600 to 2800K and observed water droplet formation. They predicted that high temperature causes ionization of N_2 and O_2 and, when this ionized air is subjected to more radiation, avalanche breakdown of air can occur.

Braun et al. (1995) have observed laser induced condensation and water drops formation by shooting self-channeling of high- peak power femtosecond laser pulses in the air. Yoshihara et al. (2007) have shown that the pulsed UV-laser irradiation of ambient air induces formation of water droplets or small ice particles in the laboratory. They also observed that [O] formed in this process quickly reacts with O_2 molecules to form O_3. Rohwetter et al. (2010) have shown that ionized filament, generated by ultra-short wave laser pulses induce water-cloud condensation in the sub-saturated atmosphere in the altitude region between 45m and 75m. A team, called terra-mobile-group (TMG), consisting of scientists from Switzerland, Germany and France, have been trying to create artificial rain by laser (Kasparian et al.2000; 2003; Mejean et al. 2006; Rohwetter al. 2010; Kasparian et al. 2012). They have done simulation experiment in laboratory cloud chamber and have observed condensation and water drops formation. They also succeeded in producing tiny water particles in moderately humid air in an altitude of 45 to 75m of the atmosphere by terawatt mobile laser. But the droplets were about a hundred times too small to fall as raindrop; instead, they remained suspended in the air.

The team feels that it is feasible to get larger droplets if the power of laser is increased to petawatt (10^{15}watts) or exawatt (10^{18}watts).

They further say that the effectiveness of this method is much easier to gauge than traditional cloud-seeding techniques and that it could provide to be a practical means of triggering rainfall". (Search Google

as 'Laser makes rain, heavily'2010). A group of Scientists from Florida University also observed water drops formation by high power laser shooting experiment. It appears from the above that laser has not yet succeeded in producing artificial rain. In this paper, a novel method is described to create artificial rain by laser.

10.2 Condensation is the Basic Need for Water Drops Formation:

IRRA Scientist Group has been successful experiments in laboratory, for 'Condensation is the basic need for water drop formation'. That condensation is the basic need for water drop formation can be understood by taking two glasses, one filled with normal water and another with ice pieces. After sometime one can observe water droplets on the outer surface of the glass which contains ice but not in the other. This is due to the condensation process that occurs around the ice glass. So, IRRA Scientist Group, proposed a laser system for this research project, to create artificial lightning by initiation of endothermic reactions, similar to natural lightning phenomena, for artificial rainmaking. As a result of these reactions, temperature falls, condensation takes place, seeding occurs and it rains in an analogous way similar to rains created by lightning. This process has been practically proved in the laboratory and atmosphere as "production of ozone and nitrogen oxides by laser filamentation". A number of scientific and practical tests have been conducted in cloud chambers and in the atmosphere to prove this hypothesis, as

"Laser induce condensation and water drops formation by Laser shooting in the laboratory cloud chambers as well as in the atmosphere".

Now the question arises that what are the conditions required for condensation.

This means that, only and only, endothermic reactions are responsible for condensations, which also produces NO (Nitrogen Oxides) and O_3 (Ozone) after shooting laser beams triggering endothermic reactions, condensation and precipitation. These tiny water drops act as a natural seeding process, due to acceleration and tribulation by wind force in the atmosphere, to form another set of raindrops with heavy rainfall as lightning rain. Endothermic reactions are responsible for condensation. Condensation is the basic need for water drop formation as above laboratory experiment, "Condensation is the Basic Need for Water Drops Formation".

This novel Rainmaking Technology can be used for white, warm clouds too which get converted into black rainy clouds for rain enhancement. As well as, water drops formation by high power Laser shooting, "Laser makes rain".

The IRRA scientist Group has already demonstrated "Innovative Rainmaking Technology" in the Laboratory cloud's chamber successfully, and results have been published in "Indian Journal of Science & Technology", http://indjst.org, vol.1, No.6 (2008).

Now, the project proposal, the design of the laser system, the Budget estimate, and the work plan are ready with IRRA Scientist Group, India. IRRA Scientist is ready for demonstration and collaboration with the Government for funding purposes for project proposal on "An Innovative Rainmaking Technology using Laser system from Aircraft to form raindrops acting as the natural seeding process due to initiation of Endothermic Reactions, for rain enhancement in large scale in the atmosphere".

In this experiment, femto-second–terawatt laser creates artificial lightning in the atmospheric cloud's regions by Air craft. These white warm clouds are converted into black rainy clouds with natural seeding for rain enhancement in the atmosphere.

A laser pulse will be sent to the cloud to initiate endothermic reactions which will create lightning phenomena, as in nature, mentioned above. For example, a German-French group has used a femtosecond–terawatt laser to obtain "Laser-assisted water condensations in the atmosphere". They have succeeded in obtaining raindrops from an altitude of 45m to 75m of the atmosphere. This novel Rainmaking Technology can be used for white, warm clouds too which get converted into black rainy clouds for rain enhancement.

**As well as, water drops formation by high power Laser shooting. "Laser makes rain" by Florida University, Scientist experimentally observed.*

**As per report, a group of European scientists working on artificial rain said in 2010, "Firing extremely powerful laser pulses through humid air can stimulate the formation of clouds, according to a team of European scientists.*

They say that the effectiveness of this method is much easier to gauge than traditional cloud-seeding techniques and that it could provide to be a practical means of triggering rainfall". (Search Google as 'Laser makes rain, heavily'2010).

In Indian ancient tradition, there is a mention in Vedshastra that "Fire arrows are sent towards the atmospheric clouds which is responsible for immediate rainfall "Our system could be a terawatt femtosecond Ti: sapphire pulse laser.

Its fundamental wavelength could be ~800nm. The pulse will have energy of ~500mJ, 120fs and repetition frequency of 10Hz. The laser pulses has to propagate with almost high peak intensity in atmospheric clouds.

It works when more than 65% humidity is present in the atmosphere. Our findings could be used by scientists and engineers to create artificial rain through a new method. The results could be of immense benefit to human being.

10.3 How will laser create artificial rain? Theory:

For creation of rain, according to well established meteorology theory (Can be found in any textbook of meteorology), steps are the following:

First (i) creation of low temperature →then (ii) condensation then (iii) seed (CCN) formation →then (iv) tiny water drops formation and rain occur in the atmosphere.

The present methodology is to send laser pulse to the cloud region of the atmosphere to create high temperature. This high temperature will break the bonds of O_2 and N_2 as follows:

$$N_2 \rightarrow N \equiv N \rightarrow N^* + N \text{ ---- } (1)$$

$$O_2 \rightarrow O = O \rightarrow O^* + O \text{ ----- } (2)$$

In this processing N and O in excited state (N^*, O^*) will be created. These excited N^* and O^* are very unstable and immediately come to the ground state through following reactions.

$$N^* + O_2 \rightarrow NO + O \quad \Delta H\ (43.2\ kcal/\ mol) \ldots\ldots \ldots\ldots\ldots (3)$$

$$O^* + O_2 + M \rightarrow O_3 + M \quad \Delta H\ (67.7\ kcal/mol) \ldots\ldots (4)$$

The occurrence of reactions 3 and 4 and formation of NO and O_3 have confirmation from NASA laboratory experiments [Sanders et al. 2003]. Formation of O_3 and NO, after laser shot, has been observed in laser experiment in atmosphere (Petit et al. 2010). The reactions 3 and 4 are *endothermic and therefore, they need heat energy (amount mentioned in brackets) which is absorbed from the cloud region. As a result, temperature of the cloud region falls (first step of rain formation is achieved and then other steps follow), condensation takes place, seeds (CCN) will be formed, and tiny water drops will be created. These tiny water drops may act as natural seeds to form another sets of rain drops. This chain process will result in rainfall. Ozone and Nitric oxide, O_3 and NO (formed in reactions 3 and 4) will undergo further reaction to form HNO_3 particles other nitrogen compounds, which will bind water molecules together to create water droplets.*

These water droplets again will act as natural seeds to form another sets of rain drops. In the atmosphere, due to turbulence, small water drops coalesce and form big rain drops. In addition, ions N_2^+ and O_2^+ and electrons formed by cosmic rays can create complex hydrated heavy positive and negative ions... $HNO_3 (H_2O)_n$ (where the value of n could be as large as 50) which can also act as seed to create rain.

In short, to create artificial rain by laser, endothermic reactions are to be generated in the cloud region. Its has-been shown earlier how much heat energy is absorbed by endothermic reactions from atmospheric clouds

(Chopkar 1993a, b; Chopkar and Chakrabarty 2008; Chakrabarty et al. 2010; Chopkar et al. 2010). The energy required to break bonds of 1 molecule of N_2 and 1 molecule of O_2, = $2.25x10^{-18}$ Joule. A laser pulse of energy 500mJ can dissociate a column of N_2 and O_2 containing ($\sim 0.5/2.25^{-18}$) ~ 10^{17}molecules which is much higher than the density in the atmosphere?

Laser can be operated from the ground as well as from an aircraft. In former case, laser pulse has to propagate to a height of ~1km (cloud height) from the ground. There will be attenuation of energy in this propagation. Kasparian et al. (2012) experimented with terawatt laser from the ground and observed tiny raindrops in an altitude of 45m to 75m of the atmosphere. To create large water droplets at higher altitudes, the group feels that laser power has to be peta-watt (10^{15}watt) or hexa-watt (10^{18}watt). If laser is operated from aircraft, then attenuation of energy will be less. In that case, laser power can reach the cloud region without much attenuation. It can also cover large area and can move to any place.

Turbulence created by the aircraft in the atmosphere can also create small water drops which would collide with each other and form big rain drops.

10.4 Methodology for Artificial Rainmaking by Laser system, on board Aircraft:

We propose a laser system of following specification: 10^{12}watt, 800nm,

500mJ, 120fs and 10Hz for this research project on board aircraft (Fig.1).

In aircraft, as shown in Fig. 2, there are two floors. In the 1st Floor of height 9', the parts shown as (1), (2) and (3) are where laser system is installed, and in the 2nd Floor of height 7' 6" parts shown in (A), (B), (C) and (D) with stair case are where power supply unit is installed. There is railing for open space shown as square dots. In the aircraft, seven Laser instruments shall be installed, 3 on right side, 3 on left side (as shown in part-1) and one on back side. Aircraft will release seven high power laser pulse in the atmospheric clouds. They will cover more than 200Km^2 area. Speed of aircraft will be ~100km/hr.

In aircraft there shall be instruments by which we have to measure atmospheric parameters such as humidity, temperature, pressure, wind velocity, wind direction, etc. on ground level as well as in the upper level of atmosphere. When the upper atmosphere is cloudy and more than 65% humidity is present in the atmosphere; the experiment can be started. After the completion of the experiment, all data are analyzed.

In this experiment, femto-second–terawatt laser creates artificial lighting in the atmosphere up to 1.2 Km to 3.2 Km altitude in the white warm cloud's regions.

These white warm clouds are converted into black rainy clouds with natural seeding for rain enhancement in the atmosphere.

A laser pulse will be release from Aircraft as multiple lightning creates in to the upper atmospheric cloud to initiate endothermic reactions which will create lightning phenomena, as in nature, mentioned above. The laser technology for this purpose, though not fully developed, yet exists. For example, a German-French group has used a femtosecond–terawatt laser to obtain "Laser-assisted water condensations in the atmosphere".

*They have succeeded in obtaining raindrops from an altitude of 45m to 75m of the atmosphere. This novel Rainmaking Technology can be used for white, warm clouds too which get converted into black rainy clouds for rain enhancement. *As well as, water drops formation by high power Laser shooting. "Laser makes rain" by Florida University, Scientist experimentally observed.*

**As per report, a group of European scientists working on artificial rain said in 2010, "Firing extremely powerful laser pulses through humid air can stimulate the formation of clouds, according to a team of European scientists.*

They say that the effectiveness of this method is much easier to gauge than traditional cloud-seeding techniques and that it could provide to be a practical means of triggering rainfall". (Search Google as

'Laser makes rain, heavily'2010). In Indian ancient tradition, there is a mention in Vedshastra that "Fire arrows are sent towards the atmospheric clouds which is responsible for prompt rainfall"

Our system could be a terawatt femtosecond Ti: sapphire pulse laser.

Its fundamental wavelength could be ~800nm. The pulse will have energy of ~500mJ, 120fs and repetition frequency of 10Hz. The laser pulse has to propagate with almost high peak intensity over a distance of ~1km.

It works when more than 65% humidity is present in the atmosphere.

Our findings could be used by scientists and engineers to create artificial rain through a new method. The results could be of immense benefit to human being.

It will initiate endothermic reactions which will create lightning phenomena and rain as in nature. These white warm clouds will be converted in to black rainy clouds with natural seeding for rain enhancement in the atmosphere.

When the upper atmosphere is cloudy and more than 65% humidity is present in the atmosphere, the experiment can be started in the field of the farmer.

We will measure atmospheric parameters such as humidity, temperature, pressure, wind velocity, wind direction, etc. on ground level as well as in the upper level.

For this purpose, a separate unit/department must be established as

"Measuring & Maintains Department". After the success of experiment, all related data, must be put it in software computer, for analysis and conclusion, with fixed perfect Laser design for maximum rainmaking / rainfall in the atmosphere.

In this way, we will use Laser system as Femto-second terawatt Laser 10^{12}*watt power, on board Aircraft. Similarly, Laser power has to be peta-watt* (10^{15}*watt) or hexa-watt* (10^{18}*watt) should be used for creating artificial lightning in the atmosphere for rainmaking from ground level which is send as Laser pulses length 2.5m to 5.0m in repetition process in upper cloud regions.*

Demonstration on project proposal of Innovative Rainmaking Technology by Laser system on board Aircraft for Rain enhancement in the Atmosphere by IRRA Scientist Group.

"Artificial Rainmaking Methodology by Laser system on board Aircraft", it's most useful for "Green revolution in the whole world for all human beings".

10.5 Discussion:

Innovative Rainmaking Technology is scientifically and practically proven as "Laser induces condensation and water drops formation in Laboratory cloud's chamber as well as in the atmosphere

However, according to Kasparian group, a laser pulse shot in the atmosphere ionizes

N_2 *and* O_2

$N_2 + hv \rightarrow N_2^+ + e^-$ ……….. (7)

$O_2 + hv \rightarrow O_2^+ + e^-$ ……… (8)

They have observed lightning phenomenon in the laboratory cloud chamber as "Laser induced condensation and water drops formation in the laboratory cloud chamber by Femtosecond –Terawatt mobile laser system". Kasparian (2012) group says that it is the ionized species N_2^+ and O_2^+ which produce rain.

But these two species are of micro size which cannot act as seeding agents. Also, N_2^+ and O_2^+ radicals are not observed by Kasparian group in laser filamentation experiments but production of O_3 and NO has been observed by them in laser filamentation experiments. Experiments of Kasparian group finds condensation and water drop formation and they say in them ionization theory that N_2^+ and O_2^+ act as seeding agent. They also say

"Mechanism of laser–induced condensation involves photo dissociation, in which photons break down atmospheric compounds in the atmosphere".

This process produces Ozone and Nitrogen oxides, which lead to the formation of Nitric acid particles that bind water molecules together to create water

droplets." But there is no seeding and condensation and water drop formation is not due to seeding. Small water drops formed by laser in the laboratory cloud chamber are due only to endothermic reactions (cooling) and this is obvious. In the atmosphere, due to acceleration and turbulence, these small sized water drops coalesce to form big rain drops. These rain drops act as a natural seeding process to form different sets of rain drops; this chain process continues with heavy rainfall.

Calculation for the energy required for dissociation is almost half of that required for ionization

The energy of a laser beam of wavelength λ is hν (ν = 1/λ and h is Planck's constant). We will shoot laser pulse in the atmosphere and dissociate (break bonds of) N_2 and O_2 as follows:

$N_2 + hv \rightarrow N^* + N$ …… (1)

$O_2 + hv \rightarrow O^* + O$ …… (2)

Bond energy of N_2 *= 226 kcal/mole.*

1 cal = 4.184 Joule, Avogadro number = $6x10^{23}$

Therefore, energy required to break 1 molecule of

$N_2 = 226x10^3x4.184/\ (6x10^{23}) = 1.58x10^{-18}$ *Jou*

Bond energy of O_2 *= 96 kcal/mole.*

Therefore, energy required to break 1 molecule of

$O_2 = 96x10^3x4.184/\ (6x10^{23}) = 0.67x10^{-18}$ *Joul*

So, the total energy required for breaking 1 molecule of N_2 *and 1 molecule of*

O_2 *will be* $(1.58x10^{-18} + 0.67x10^{-18})$ = <u>$2.25x10^{-18}$ *Joule.*</u>

When a laser pulse is shot in the atmosphere, it may ionize N_2 *and* O_2 *as follows:*

$N2 + hv \rightarrow N_2^+ + e$......... (3)

$O2 + hv \rightarrow O_2^+ + e$.........(4)

Ionizing potential of N_2 *= 15.58 ev = 2.49x10-*^{18}*Joule*

Ionizing potential of O_2 *= 12.2 ev =* $1.95x10^{-18}$*Joule*

So the total energy required ionizing 1 molecule of N_2 *and 1 molecule of* O_2 *is* $2.49x10^{-18}$ *Joule +*$1.95x10^{-18}$*Joule =* <u>$4.44x10^{-18}$ *Joule.*</u>

The above calculation shows that the energy required dissociating 1 molecule of N_2 *and 1molecule of* O_2 *is about half of that required to ionize them.*

The energy required for dissociation is almost half of that required for ionization.

That means energy is first used up for dissociation, then the remaining energy (Which may not be sufficient for ionization of N_2 *and* O_2*) is delivered for ionization of* N_2 *and* O_2*. Hence dissociation takes place and not ionization. The Kasparian group does not talk about dissociation. It is not only near the IR laser system,*

Yoshihara, et al. (2007), have discussed in their paper the possibility of creating artificial rain by using UV lasers. Our methodology is to send laser pulses to cloud regions to break bonds of O_2 and N_2 (by reactions 1 and 2), create endothermic reactions and condensation (by reactions 3 and 4) and produce rain in the similar way as in lightning. There is attenuation of energy in operating the laser through Aircraft. Kasparian's group suggests an increase of laser power to petawatt(10^{15}watt) or exawatt (10^{18} watt) to create large water droplets. We will operate from an aircraft in the same way as spraying chemicals from Aircraft. A laser pulses of energy 500mJ is capable of dissociating a column of N_2 and O_2 containing ($\sim 0.5/2.25^{-18}$) ~ 10^{17} molecules which is much higher than the density in the atmosphere.

10.6 Additional Uses:

This method can be used for rain harvesting by "rain drain". When huge clouds are present above a lake or dam, a laser beam can be shot into the cloud region; then with blast of clouds heavy rainfall will occur to fill the lake or dam for future use of water. This method can also be used to reduce pollution of the atmosphere by spraying artificial rain on the polluted city.

Another use of this method is to stop excess rainfall. Low intensity laser pulse shot into the cloud region will evaporate the clouds from the excess rainfall area.

This method can also be used to drive away the rain cloud from the region where rain is not needed.

10.7 Acknowledgement:

We express our sincere thanks to scientists: Prof D.K. Chakarbarty (PRL Ahmadabad Prof Mamata R. Lanjewar (RTM University Nagpur), Prof A. P. Deshpande (Principal, Science College), Prof. Padmanabhan Murthy (J.N. University, New Delhi), Prof Umesh Kulshetra (J.N. University, New Delhi), Prof K. S. Korgaokar (Pune University, Pune),

Dr. A.L. Agarwal (NEERI Nagpur), Dr. Nitin Saraf (B.D. Engineering College, Sewagram), Dr. K. M. Kharate, (GM Engineering College &

Research Centre, Sheogao), Dr. Mrs. Sumanlatha Pandey, Mr. Shyam Ujjaninkar (Engineer), Mr. Rajesh Iyengar (P. N. College, Nanded) for their valuable suggestions and guidance for project proposal.

10.8 Conclusion:

"Artificial rainmaking in large scale by laser system which initiate endothermic reactions, as similar natural lightning phenomenon, onboard Aircraft in the atmosphere" *It is shown in this project proposal that by initiating endothermic reactions in the cloud region of the atmosphere by high power laser, artificial rain can be created. Laser may have the following specification: 10^{12}watt, 800nm, 500mJ, 120fs and 10Hz for operation onboard Aircraft. This method is economical (one time investment), harmless, eco-friendly and can be switched on and off when desired. It can be used at any place and at any time. It can also be used to fill lake or dam for storing rain water for future use (rain harvesting), to reduce pollution by spraying artificial rain on the polluted city, to stop rain in the region where it is not wanted or where rainfall is in abundance. It may not be out of place to state here that in the holy Hindu book "Mahabharata", there is a mention that by firing arrows in the atmosphere, rain was created by Arjun to quench the thirst of God "Bisham".*

Artificial rain making methods help to increase the green life and oxygen and decreasing the pollution. Thus, these methods play major role in reducing drought and increasing the quantity of drinking water in future.

10.9 References:

* Braun, A., Korn, G., Liu, X., Du, D., Squier, J., and Mourou, G., 1995. "Self-channeling of high- peak power femtosecond laser pulses in air." Opt. Lett., vol. 20, pp. 73-75.

*Carls, J. C. and Brock, J. R., 1987. "Explosion of a water droplet by pulsed laser heating."

Aerosol Sci. Technol., vol. 7, pp. 79-90.

*Chopkar, S. K. and Chakrabarty, D. K., 2008. "Artificial rainmaking system in a way of natural phenomena." Indian J. Sci. Technol., vol. 1, pp. 1-5.

Available: *http://www.indjst.org*

*Chakrabarty, D. K, S. K. Chopkar and N. N. Purkait, "Femtosecond terawatt laser system to produce artificial rain" presented at 4th international conference on computer and devices for communication, CODEC 2009 at Kolkata in December 2009 and published in IEEE X-pore digital library, INSPEC Accession No. 11136783, Print ISBN

978-1-4244507329, February 2010.

*Chopkar, S. K., "Effect of endothermic reactions associated with lightning on atmospheric chemistry" Indian J. Radio Space Phys.22, 128-131, 1993a.

*Chopkar, S. K., *American Meteorological Society in Meteorological & Geo-astrophysical*

Abstracts,44, October 1993 No.10 (44.10-555) 1993b.

*Chopkar, S. K., D. K. Chakrabarty, S. M. Sonbawane, R. L. Bakal, P. S. Chopkar, P. Hariom and J. S. Pimplkhute, International J. of Meteorology, 35(355), 363-370, 2010.Golde, R. H., Lightning, vol. 1, Physics of Lightning, Academic Press, London, 1977.

*Kasparian, J., R. Sauerbrey and S. L. Chin, The critical laser intensity of self-guided light filaments in air, Appl. Phys. B 71, 877-879, 2000.

* Kasparian, J., M. Rodriguez, G. Mejean, J. Yu, E. Salmon, H. Wille, R. Bourayou, S. Frey, Y. –B. Andre, A. Mysyrowicz, R. Sauerbrey, J. –P. Wolf and L. Woste, White-light filaments for atmospheric analysis, Science, 301, 61-64, 2003.

*Kasparian, J, P. Rohwetter, L. Wöste and J. –P. Wolf, Laser-assisted water condensation in the atmosphere: a step towards modulating precipitation,

Journal of Physics D: Applied Physics,45 (29), 2012.

*Mejean G., Ackermann R., Kasparian J., Salmon E., Yu J., Wolf J. -P., Rethmeier K.,

Kalkner W., Rohwetter P., Stelmaszczyk K. and Woste L., (2006) Improved laser triggering and guiding of megavolt discharges with dual fs-ns pulses, App. Phys. Letts.,88, 021101

*Petit, Y., S. Henin, J. Kasparian and J. –P Wolf, Production of ozone and nitrogen oxides by laser filamentation, Appl. Phys. Lett. 97, 021108, 2010.

*Rohwetter, P., J. Kasparian, K. Stelmaszczyk. Z. Hao, S. Henin, N., Lascoux, W. M.

Nakaema, Y. Petit, M. Queisser, R. Salame, E. Salmon, L. Woste and J. –P. Wolf, Laser-induced water condensation in air, doi: 10.1038/nphoton.2010.115, 2010.

*Sander, S. P., R. R. Friedl, D. M. Golden, M. J. Kurylo, R. E. Huie, V. L. Orkin,

G. K. Moortgat. Ravishankar, C. E. Kolb, M. J. Molina and B. J. Finlayson-Pitts, Chemical kinetics an photo-chemical data for use in atmospheric studies, NASA JPL publication, pp. 02-25, 2003.

*Yoshihara, K., Y. Takatori, K. Miyazaki and Y. Kajit, Ultraviolet light-induced water- droplet formation from wet ambient air, Proc. Jpn. Acad. Sci. B 83, 320.

By- IRRA Scientists Group

10.11: Figures for Demonstration on Rainmaking:

Demonstration on Innovative Rainmaking Technology by Laser system onboard Aircraft

Fig. No. 1

Fig. No. 1 - For Rainmaking by Laser using Aircraft in the atmosphere.

Figure No.2: "Inner Deign of Aircraft with Laser system and *Position of Laser system in Aircraft"*

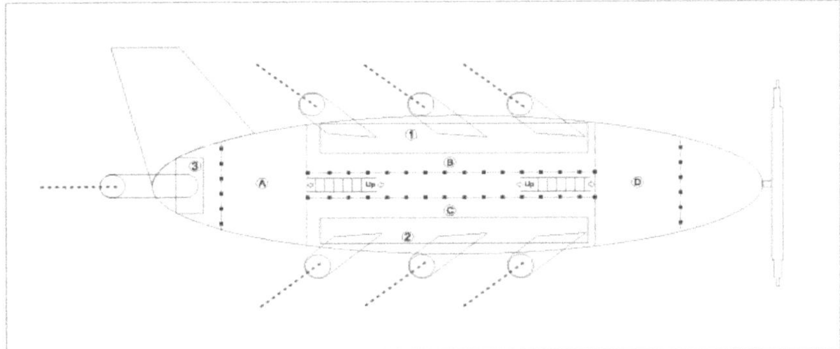

Fig.No.2: "Position of Laser system in Aircraft"

*Note: * In figure No. 2, on first floor, height 9', shown parts (1), (2) and (3) are high power Laser release system which creates multiple artificial Lightning in the upper atmospheric clouds directly by Laser system ......*

** In figureNo.2, on second floor, height 7' 6", shown parts (A),(B),(C) and (D) with stair case are high power*

generation system for high power Laser system, with railing for open space as shown square dots ....

10.12: Budget For Rainmaking Project Proposal for Three Years in The Atmosphere:

"Artificial rainmaking by using high power laser which initiate endothermic reactions, as

Similar natural lightning phenomenon, onboard Aircraft in the atmosphere"

Grand Total Estimate for Rainmaking Project proposal

(A+B+C+D+E+F+G) - 230 Cr in Indian Rs/ USA 28.75 million Dollar)

Sr.	Items	Budget for Rainmaking Project			
		1st Year	2nd Year	3rd Year	Total
A	Salary of Staff	02,03,76,000	02,31,50,000	02,54,96,000	6,90,22,000
B	Equipment's purchasing and maintains per years	155,79,00,000	10,00,00,000	10,00,00,000	(155,79,00,000/- +20,00,00,000) = 175,79,00,000
C	Consumable Items	03,05,00,000	03,05,00,000	03,05,00,000	9,15,00,000
D	Travel Expenses	05,00,00,000	04,0000,000	04,00,00,000	13,00,00,000
E	*Patent Filling Expenses*	01,20,00,000	01,20,00,000	01,20,00,000	3,60,00,000
F	Contingencies	04,10,00,000	04,10,00,000	04,10,00,000	12,30,00,000
G	Overhead Expenses	03,10,00,000	03,10,00,000	03,10,00,000	9,30,00,000
	Total	174,17,76,000	27,66,50,000	27,89,96,000	230,02,22,000

IRRA Scientist Group successful in above rainmaking technology, as scientifically & practically in Laboratory cloud chamber as well as in atmosphere up to 75M altitude, (As per demonstration on laboratory cloud chamber by IRRA scientist Group and References), then we will demonstrated/experiment from special design Aircraft (Shown in Fig.No.2) in the atmosphere for large scale for three years, project proposal estimates cost 230 Cr./USA dollar 28.75 million.

Demonstration on project proposal of Innovative Rainmaking Technology by high power Laser system for Rain enhancement in the Atmosphere by IRRA Scientist Group.

"Artificial rainmaking by using high power laser which initiate endothermic reactions, as in natural lightning phenomenon, onboard Aircraft in the atmosphere.

A) Budget for Salaries of Staff:

SN	Designation	Pay Scale	Monthly	Staff	Budget for Salaries of Staff			
					1st Year	2nd year	3rd Year	Total
1	Senior Scientist	Class 1	160000	2	3840000	4320000	4780000	1,29,40,000
2	Assistant Scientist	Class 2	120000	3	4320000	4780000	5160000	1,42,60,000
3	Senior Pilot	Class 2	180000	2	4320000	4780000	5160000	1,42,60,000
4	Junior Scientist	Class 3	80000	4	3840000	4320000	4640000	1,28,00,000
5	Lab Helper	Class 4	35000	3	1260000	1540000	1860000	46,60,000
6	Skilled Labor	Class 4	25000	5	1500000	1770000	2000000	52,70,000
7	Non- Skilled Labor	Class-4	18000	6	1296000	1640000	1896000	48,32,000
	Total				20376000	23150000	25496000	6,90,22,000

B) Budget of Equipment Purchase:

SN	Particular	Model	Make	Estimate Cost (In Lakhs)
1	High power laser Instrument with Transportation (Femto-second high power Laser Instrument)	For Seven Numbers with multiple High power laser release instruments. (87 Lakhs/per x 7 Nos)	Imported	06,09

2	Cost of special Aircraft Design as per requirement as shown in Fig. No.2	Special design as shown in Fig.No.2	Imported	87,00
3	High power supply point /Fuel Materials for high power Laser system in Aircraft instruments.	Special design as shown in Fig.No.2	Indigenous	30,00
4	Precipitation & Atmospheric parameters measuring instrument.		Indigenous	14,00
5	Natural Lighting Protector Unit, as Conductor with earthling for Laser Instrument & Air Craft		Imported	02,20
6	*Construction cost for Equipment & Camp/Mobile Van Two Numbers*		Indigenous	16,50
	Total			155,79,00,000

C) Budget for Consumable Item:

Budget of Consumable Item					
Sr.	Consumable Items	1st Year	2nd Year	3rd Year	Total (Lakhs)
1	high power supply unit for laser system	00,50	00,50	00,50	01,50
2	*Fuel materials*	00,75	00,75	00,75	02,25
3	Stationary & another misalliance material	00,60	00,60	00,60	01,80
4	Small instruments / Equipment for Laboratory uses and other purpose.	00,80	00,80	00,80	02,40
5	Chemicals for testing or Analysis purpose for laboratory	00,40	00,40	00,40	01,20

	Total	03,0 5	03,05	03,0 5	09,15

D) *Budget for Travel Expenses:*

Budget of Travel Expenses					
SN	Description	1st Year	2nd Year	3rd Year	Total (Lakhs)
1	*Conference for advance technology for discussion and suggestion for inviter*	*02,20*	*02,20*	*02,20*	*06,60*
2	*Visit of top scientists to project for various Suggestion for T.A.&D.A.*	*01,00*	*01,30*	*01,30*	*03,90*
3	*Vehicle purchase for transport Two van and two motor cycles and Maintenance*	*01,80*	*00,50*	*00,50*	*02,80*
	Total	*0500*	*0400*	*0400*	*13,00*

E) *Budget for Patent Filling Expenses*

Budget of Patent Filling Expenses					
SN	Description	1st Year	2nd Year	3rd Year	Total (Lakhs)
1	*National & International Patent file expenditure*	*00,40*	*00,40*	*00,40*	*01,45*
2	*Consultant fees / Expert guidance Service charges*	*00,80*	*00,80*	*00,80*	*02,40*
	Total	*01,20*	*01,20*	*01,20*	*03,60*

F) Budget for Contagions

Budget for contagions					
SN	Description	1st Year	2nd Year	3rd Year	Total (Lakhs)
1	Contagions	*04,10*	*04,10*	*04,10*	*1230*
	Total	*04,10*	*04,10*	*04,10*	*12,30*

G) Budget for Over Head:

Budget for Over Head					
SN	Description	1st Year	2nd Year	3rd Year	Total (Lakhs)
1	*Over Head*	*03,10*	*03,10*	*03,10*	*09,30*
	Total	*03,10*	*03,10*	*03,10*	*09,30*

By-

IRRA Scientist Group, India.

10.13: Work-Plan for Three Years:

"Artificial rainmaking by using high power laser which initiate endothermic reactions, as in natural lightning phenomenon, onboard Aircraft in the atmosphere"

Three Years Work Planning

First Year		
1-4 Months	*5-8 Months*	*9-12 Months*
Appointment of scientists, piolet, skilled and non-skilled workers. Select and confirm project active team for project demonstration.	*Inviting experts from the concerned area for table discussion on the demo of the project with consultant expert scientist etc.*	*Demand for different instruments to different tenders for special aircraft, high power Laser instruments, & others.....*
Select Airport site for special Aircraft design for accessories, assembly erection with laser instruments in special Aircraft	*Organization of two days national & international conference on the demo of the project with discussion to varies expert scientist, specializations*	*Comparative statements of received quotations for the different instruments &work order to reputed company etc.*
All facilities provide for scientists, gusts & workers on project for demonstration arranged on site as temporary base as boarding etc.	*As per expert suggestion, comments fallow their important direction in project technology as per IRRA Scientists decision......*	*Purchase work order of proper instrument to proper manufacturing company & suppliers etc. .....*

Second Year		
1-4 Months	*5-8 Months*	*9-12 Months*
All instrument & required accessories collected on project site for erecting special Aircraft with high power laser instruments.	*carrying experiments on the selected site and recording the observations as per the direction given by the experts*	*Repetition experiment in atmospheric clouds with improving technology & records observation & Interpretation of the data with comparison....*
Accessories of high-power laser instruments connected with supply high power in special Aircraft completed as per Fig. No.2 shows etc.	*Inviting experts from the concerned areas for table discussion on the observations taken……*	*Compilation of the data in computer software and commercialization of the same for the benefit of society.*
Trial for testing all connection for high power supply for high power laser instruments	*Analysis of the observations taken in demonstration of rainmaking project. And improved in technology .....*	*Publication of the result in the reputed national and international journals*
Third Year		
1-4 Months	*5-8 Months*	*9-12 Months*
Analysis of the observations & confirm proper solution for maximum rainfall	*Rainmaking technology use for commercial to benefit society for all human being in whole world*	*Filing the patent for the artificial rainmaking system Analysis of the observations*
Interpretation of the data in software computer for future beneficial to society	*IRRA scientists aim fulfil as "Green revolution in the whole world for all human beings "*	*Appeal to the Indian government and other countries to utilize this project for the benefit of Humankind.*
Innovative rainmaking technology use in education syllabus for research study	*In press conference, recommended "Innovative rainmaking technology "for research study for beneficial to all society …….*	*Research study most useful for provides as water and food for all human beings in whole world as "Green revolution "……….*

14.0: Request to Encourage IRRA Scientists Research Activities:

IRRA Scientists have developed an "Innovative Rainmaking Technology " which is scientifically and practically proven in Laboratory Clouds chamber as well as in the atmosphere up to 75m altitude as Ref. "Laser induces condensation and water drops formation in laboratory clouds chamber".

Now, IRRA scientists prepared new project proposal with design, budget & estimate, work plan etc. on "Artificial rainmaking by laser system which initiate endothermic reactions, as similar natural lightning phenomenon, onboard Aircraft in the atmosphere" which is useful for demonstrations in the atmosphere by any Government /Organization /Company. Hearty request by IRRA Scientist to encourage IRRA Scientist's research activities for "Green revolution in the whole world for all human beings", please do the needful & consider.

"God's Gift from Shri Sadguru Ramchandra Namamahay Hari Om!"

Now days, IRRA scientist group are struggling with funds for research activities.

Your little help may be most useful for "Green revolution in the whole world for all human being". Your co-operation is solicited.

Your help may be most useful for all human being, in the whole world for green revolution.

Visit website- *http://www.irraindia.org* E skc.arr@rediffmail.com, E-irra.scientistgroup@gmail.com

Please visit our YouTube Channel search as" Novel Technology for Artificial Rainmaking'

Click here :- https://youtu.be/UYfuH9fAmUs

*Please search on Google as S.K. Chopkar or "Rain Enhancement by

Endothermic reactions" or visit our web site http://www.irraindia.org

Bank Account Details: -

Bank Name-CENTRAL BANK OF INDIA, MIDC Area Wardha,

SEWAGRAM ROAD—442006, India

Bank A/c Name: - Innovative Rainmaking Research Association, Sewagram- 442102, Dhanvantri Nager-30, Wardha

Bank A/c No. 3806623156, IFSC Code No.: - CBIN0282189,

15.0: "Request Letter"

"Schidanand Pandurang HriOm"

"GREEN REVOLUTION"

INNOVATIVE RAINMAKING RESEARCH ASSOCIATION (IRRA), INDIA

Prof.D.K. Chakrabarty (Scientific Director) IRRA Scientist Group, India 40/178, GF CR Park, NEW DELHI -110 019 *Mob. No. +91-990121979, 09327020993* *E - dkchakrabarty@rediffmail.com* E- *irra.scientistgroup@gmail.com*	 	Dr. S.K. Chopkar (Director of IRRA scientists) Dhanvantari Nagar- 30, SEVAGRAM-*442 102* *Dist. Wardha, Maharashtra, India* Mob. No.+91-9420445108 E-mail- skc.arr@rediffmail.com. *website http://www.irraindia.org*

Date: 16.03.2023

Respected Sir,

Greetings from IRRA Scientists!

Subject: - To encourage research activities of IRRA Scientists in the Field of Innovative Rainmaking Technology"

Respected Sir,

Innovative Rainmaking Technology is scientifically and practically proven in laboratory cloud chambers, as well as in the atmosphere. This includes

"Laser-induced condensation and formation of water drops in laboratory cloud chamber as well as in the atmosphere up to 75m altitude."

IRRA Scientists Group would like to information about our organization "Innovative Rainmaking Research Association (IRRA Scientists), India on research activities of Novel Rainmaking Technology from last 37years, which is scientifically & practically proven. Our organization are working in a rural area as a social, scientific, for social welfare. Our organization register under GOVERNMENT OF INDIA MINISTRY OF CORPORATE AFFAIRS Central Registration Centre CERTIFICATE OF INCORPORATION,

Section -8. Our research activities are on a no-profit and no-loss basis. We work for the welfare of society. We have no source of income; no funding from anyone. Our scientists work here voluntarily during their free time. We manage by contributing money from our own pockets.

We are delighted to provide information on the research activities of the IRRA Scientists group, Sewagram, India who have developed for their excellent research, an innovative method of artificial rain forming. This invention is very much valuable for mankind. This is a great contribution to atmospheric Chemistry. We strongly recommend the IRRA Scientists group, Sewagram India for encouraging research activities.

IRRA Group developed a challenging research work "Innovative rainmaking technology by Laser system similar to natural lightning phenomena in the atmosphere ". This technology is scientifically and practically proven in laboratory cloud chambers, as well as in the atmosphere. This includes as result "Laser-induced condensation and formation of water drops in laboratory cloud chamber as well as in the atmosphere". IRRA Scientists aim for a "Green revolution in the whole World for all human beings". Please let us know if you have any suggestions /comments contact us! Without any hesitation to IRRA Scientists.....

IRRA Scientists would like to collaboration / partnership with your Government of Environment Science Department / Institutions /Foundation / Company for "Green revolution in the world for all human beings" IRRA Scientists very happy to work with them on "Innovative rainmaking technology" for artificial rainmaking in large scale as natural lightning phenomena in the atmosphere.

IRRA Scientists propose a laser design system of specification: 10^{12}watt, 800nm, 500mJ, 120fs and 10Hz to create multiple artificial lightening (as shown Fig.) for rainmaking on a large scale by initiating endothermic reactions as similar natural lightning phenomena on board Aircraft, in the atmosphere. Laser design creates multiple lightning in the atmosphere on board Aircraft, Laser design, estimate budget, work plan etc. are ready with IRRA Scientists for demonstration, , it's most useful for "Artificial rainmaking in large scale by using high power laser which initiates endothermic reactions, as a similar natural lightning phenomenon, onboard Aircraft in the atmosphere" It's most useful for mankind's providing food & water for all human beings "Green revolution in the world for all human beings". IRRA Scientists are ready for discussion on "Innovative rainmaking technology for demonstration".

Now, IRRA scientists have prepared a review proposal on "Innovative Rainmaking Technology for Artificial Rainmaking in large scale by Laser system onboard Air craft" which is enclosed here for your kind consideration to encourage IRRA Scientist's research activities for demonstration by any Government/Organization/Company. Sir, it is our hearty request to you and your team to consider our review proposal IRRA Group has developed a challenging research work "Innovative rainmaking technology by Laser system similar to natural lightning phenomena in the atmosphere ". This technology is practically proved in laboratory cloud chambers too, as well as in the atmosphere. This includes

"Laser-induced condensation and formation of water drops in laboratory cloud chamber as well as in the atmosphere".

Now the question arises that what are the conditions required for condensation? This means that only, endothermic reactions are responsible for condensations, which also produces NO (Nitrogen

Oxides) and O_3 (Ozone) after shooting laser beams it can endothermic reactions, condensation and precipitation. These tiny water drops act as a natural seeding process, due to acceleration and tribulation by wind force in the atmosphere, to form set of raindrops with heavy rainfall as lightning rain. Endothermic reactions are responsible for condensation. Condensation is the basic need for water drop formation. This novel Rainmaking Technology can be used for white warm clouds too which gets converted into black rainy clouds.

As per reports a group of European scientists working on artificial rain said in 2010 that "Firing extremely powerful laser pulses through humid air can stimulate the formation of clouds, according to a team of European scientists. They say that the effectiveness of this method is much easier to gauge than traditional cloud-seeding techniques and that it could provide to be a practical means of triggering rainfall". (Search Google as 'Laser makes rain, heavily' 2010). Florida University scientists too have experimentally observed that "Laser makes rain ".

This lightning phenomenon created through artificial lightning by plasma laser pulse or laser system can produce rain in the atmosphere has been practically proved as "Laser-induced water condensation in air" is mentioned in the literature. Scientists have succeeded in obtaining raindrops from an altitude of 75m of the atmosphere by terawatt mobile laser. "IRRA Scientist Group propose laser system of specification: 10^{12}watt, 800nm, 500mJ, 120fs and 10Hz for this research project".

It works when more than 65% humidity is present in the atmosphere. Our findings could be used by scientists and engineers to create artificial rain as a new method. The results could be of immense benefit to human being as well as eco-friendly and cost effective which is the need of the hour.

Following are the list of patents file by IRRA group, India

"National and International Patent File"

** International Australian Government, IP Australia Certificate of Grants Innovation*

Patent number: 2020101897 (2021).

**International Publication Patent No: -WO/2008/062441 (2007) and Revised International*

Application No.PCT/IN2017/000105 (2017).

**Indian Patent Number: - 238000 (2007) and Revised Indian Patent File Number 201721008920 (2017).*

IRRA Group has published number of research papers in National and International Journals (www.irraindia.org). "The International Journal of Meteorology (U.K.), Volume 35, number 355, November 2010 issue (www.ijmet.org). Very recently IRRA Group has been awarded "International Scientist Award". IRRA Group's aim is to produce "Green revolution in the whole world for all human being". IRRA Scientist Group has presented research paper in

National & International Conference. This idea is a gift to us from God for Green revolution in the whole world for all human beings.

Now the IRRA Scientists group wants to demonstrate the project. Your helpful suggestion with scientific comments is highly appreciated. Your help may be useful for "Green revolution in the whole world for all human beings". Please inform if any further formality needs to be done. Your able guidance is solicited.

Please refer attached project proposal file. Sir, please transfer research project proposal files to Government /Organization /companies who are interested for demonstration!

Please let us know, if you have any suggestion / queries / questions, contact! without any hesitation to IRRA Scientists, project proposal with Laser design, estimate bugged, work plan etc. are ready for demonstration on rainmaking in large scale in the atmosphere with IRRA Scientists Group.

IRRA Scientists are ready for discussion on research project proposal "Innovative Rainmaking Technology" and are eager to demonstrate it wherever needed Look forward to your kind consideration.

IRRA Scientists aim for artificial rain at any place, at any time, as per human needs for the green revolution of the world for all human beings. IRRA Scientists would like to be ready for discussion on "Innovative rainmaking technology" for artificial rainmaking on a large scale as similar natural lightning phenomena in the atmosphere, your suggestions/comments are welcome.

Please do the needful for a demonstration on "Innovative rainmaking technology" for artificial rainmaking on a large scale as natural lightning phenomena in the atmosphere, also, it's used for reducing pollution on the pollute area, pollution dust particles come down with artificial rain drops on the ground.

IRRA Scientists would like to collaboration / partnership with your Gov. of environment scientists / Research Institutions /Company for "Green revolution in the world for all human beings"

IRRA Scientist Group, India.

Sewagram, Dhanvantri Nager-30, Dist-Wardha-442102, Maharashtra, India

Mob. +91 9420445108, Email- skc.arr@rediffmail.com,

E- skchopkar@irraindia.org, E- irra.scientistgroup@gmail.com

Please visit our YouTube Channel and search "Novel Technology for Artificial Rainmaking".

Click here: - https://youtu.be/UYfuH9fAmUs

More details can be found in Google Search: as "Artificial rainmaking by endothermic reactions" or S.K. Chopkar or on the website: http://www.irraindia.org.

16.0: Research Paper Publish in National and International Journal by IRRA Scientist Group"

INDEX			
SR NO	*PAPER TITLE & AUTHOR NAME*	*PUBLISHED IN JOURNAL & DISCRIPTION*	*YEAR OF PUBLICATION*
01	*"Effect of endothermic reactions associatedwith*	*Indian Journal of Radio & Space Physics, Vol22, April* *1993, By: -S.k. Chopkar.*	*1993*

	lightning on atmospheric chemistry" *(By:-- S.K.Chopkar)*		
02	*"Artificial rainmaking system in a wayof natural phenomena"* *(By S.K..Chopkar & D.k. Chakarbarty)*	*"Indian Journal of Science and Technology"* (http://www.indjst.org) ,Vol.1 No 6 (Nov. 2008)	*2008*
03	*"Femtosecond terawatt Laser produce artificial rain."By:-D. K. Chakrabarty*	*Research Paper presentation in Calcutta Conference by Dr.D.K. Chakrabarty*	*2009*
04	*"Artificial rainmaking by Laser system"* *(By S.K..Chopkar & D. K. Chakarbarty)*	*The International Journal of Meteorology* www.ijmet.org,Volume35,Number355,November2010	*2010*
05	*"Economical and non-pollute system for artificial rainmaking by laser pulse in a way of natural phenomena"* *(By S.k.Chopk & S.Sonbawane)*	*Indian Journal of Innovations Dev., Vol. 1, No. 3 (Mar 2012), ISSN* 2277 – 5390	*2012*

06	*"CLOUD FORMATION & ATMOSPHERIC RAINMAKING BY ENDOTHERMIC REACTION DUE TO PLASMA LASER & UV RADIATION INTHE ATMOSPHERE"* *(By: -S.K. Chopkar& D.k. Chakarbarty)*	International Journal of Information Technology and Business Management29th Jan 2014. Vol.21 No.1 © 2012-2014 JITBM & ARF. All rights reserved *www.jitbm.com*	*2014*
07	*"Condensation and Water Drops Formationdue to Endothermic Reaction by Laser Pulse through Natural Lighting Phenomena in the atmosphere"* *(By:- S.K.Chopkar& D.k.chakarbarty)*	*International Journal of Managerial Studies andResearch (IJMSR)* *Volume 2, Issue 10, November 2014, PP 148-155* *ISSN 2349-0330 (Print) & ISSN 2349-0349 (Online) www.arcjournals.org*	*2014*
08	*"New Phenomena for*	International Research Journal of Computer Science (IRJCS) ISSN: 2393-9842	*2014*

	Condensation and Water Drops Formation by Laser Pulse Due To Endothermic Reaction in the atmosphere" (By:- Shivshankar K. Chopkar1 , D.K. Chakrabarty),	Volume 1 Issue 2 (October 2014) www.irjcs.com	

09	*"Artificial Rainmaking By Laser system" (By:- S.K.Chopkar)*	*Research Paper presentation in "International Environment conference, at Venkatesh University* *, Triputi. By: -S.K. Chopkar.*	*2016*
10	*Artificial Rainmaking by using high power laser initiation endothermic reaction, in a wayof Natural Lightning Phenomena through Air* *Craft, in a large scale, in the atmosphere" (By: -S.K. Chopkar& D.K. Chakarbarty)*	*Journal of Business Management and Economics 5:07 July (2017).*	*2017*
11	*"Novel Technology for Artificial Rainmakingby Laser System through Air Craft in the Atmospheric Clouds Initiation Endothermic Reactions, in A Way of Natural Lightning*	"Noble International Journal of Scientific Research" ISSN(e): 2521-0246 ISSN(p): 2523-0573 Vol. 02, No. 12, pp: 90-97, 2018 Published by Noble Academic Publisher URL:http://napublisher.org/?ic=journals&id=2	*2018*

	Phenomena, in the Atmosphere" *(By: -S.K. Chopkar &D.K. Chakarbarty)*		
12	*"To Create Artificial Lightning by Laser Systemthrough Air Craft in the Atmospheric Clouds Initiation Endothermic Reaction, In A Way of Natural Lightning Phenomena for Artificial Rainmaking, In the Atmosphere"* *(By:-S.K.Chopkar& D.k.chakarbarty)*	Sumerianz Journal of Scientific Research, 2018, Vol.1, No. 2, pp. 52-57 ISSN(e): 2617-6955, ISSN(p): 2617-765X Website: https://www.sumerianz.com © Sumerianz Publication	*2018*
13	*"Innovative Rainmaking Technology By Laser system in the atmosphere"* *(By: -S.K. Chopkar)*	*Research paper presentation in "International Environment conference "Held by Sahara environment Group, at Dhakala, Morocco country.By:-S.K.Chopkar*	*2019*
14	*"Innovative Rainmaking Technology by Laser System Initiation Endothermic Reactions in a Way of Natural Lightning Phenomena in the Atmosphere"* *(By:-S.K.Chopkar& D.k.chakarbarty)*	*" International journal of Atmospheric and oceanicscience"*	*2020*
15	"INNOVATION RAIN ENHANCEMENT TECHNOLOGY BY LASER SYSTEM, PRACTICALLY	IMPACT: International Journal of Research in Engineering & Technology ISSN (P): 2347–4599; ISSN(E): 2321–8843 Vol. 9, Issue 1, Jan 2019, 29–36 © Impact Journals	2021

	PROVED IN LABORATORY CLOUD CHAMBER, ALSO IN THE ATMOSPHERE” *(By:-S.K.Chopkar& D.k.chakarbarty)*		
16	*“Innovative Rainmaking Technology byLaser System Initiation Endothermic Reactions in a Way of Natural Lightning Phenomena in the Atmosphere”* *(By:-S.K.Chopkar& D.k.chakarbarty)*	*International Journal of Economy, Energy and Environment 2021; 6(3): 67-70 http://www.sciencepublishinggroup.com/j/ijeee doi: 10.11648/j.ijeee.20210603.11 ISSN: 2575- 5013 (Print); ISSN: 2575-5021 (Online)*	*2021*

17.0: Research Activities and Achievement on Rainmaking Technology by IRRA Scientist

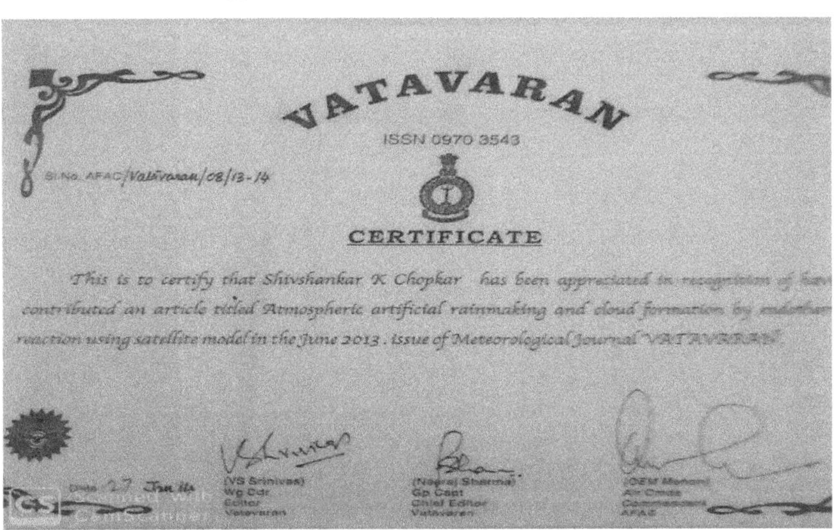

VATAVARAN

ISSN 0970 3543

Sl.No. AFAC/Vatavaran/08/13-14

CERTIFICATE

This is to certify that Shivshankar K Chopkar has been appreciated

contributed an article titled Atmospheric artificial rainmaking and cloud formation by

reaction using satellite model in the June 2013 . issue of Meteorological Journal

Date 27 Jun 14

(VS Srinivas)
Wg Cdr
Editor
Vatavaran

Gp Capt
Chief Editor
Vatavaran

Air Cmde

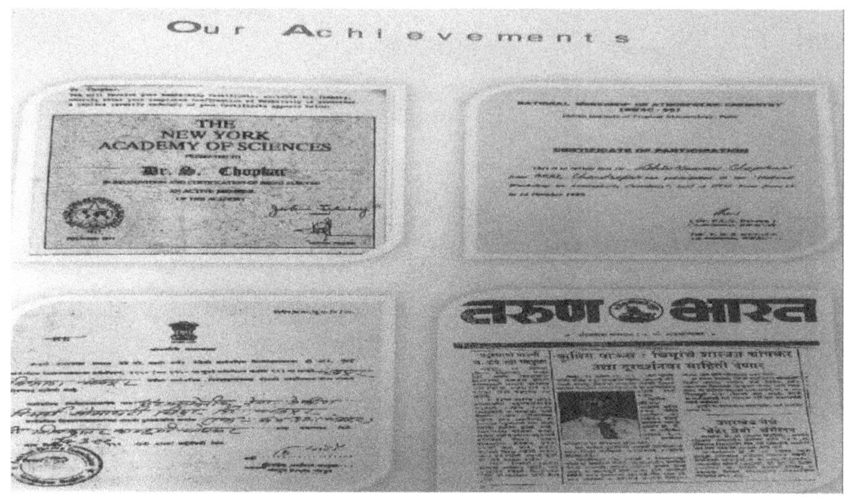

18.0: "Achievements by IRRA Scientist"

Research Paper Presented in National and International Conference

By IRRA Scientist Group, India. (Morocco International conference 27 & 28 Sept-2018)

"Novel technology for artificial rain making by laser system in a way of natural lightning phenomena"

DAKHALA, MOROCCO INTERNATIONL CONFERENCE:

Research paper "Novel Technology for Artificial Rainmaking by Laser system in a way of Natural Lightning Phenomena" presented in International Environment Conference @ Dakhla, Morocco country, sept.2018 by S. K. Chopkar, IRRA Scientist Group, India.

Certificate from Sahara Scientist Summit Dakhla, Morocco, Country

CERTIFICATE OF MERIT
for contribution to the success of the

3RD SAHARA
SCIENTISTS SUMMIT (3S)

Held in Dakhla, Morocco, 27-28th September 2018

The Organising Committee hereby certifies that

Mr/ Mme

Dr. Shivshankar K. Chopkar

participated in the scientific work of the given Summit

Prof.Dr.Jozsef STEIER
Head of 3S Org Committee

Amar Ahmed BABA
President and Chair of FAID

Dakhla, Morocco, 2018-09-28

18.1 Indian Institute of Tropical Meteorology (IITM), PUNE CONFERENCE ON ATMOSPHERIC CHEMISTRY:

Research Paper "Rainmaking Technology by Laser System Initiations Endothermic

Reactions, condensations, Precipitations, Raindrops which act as natural seeding

Process, Enhancing for Rainfall in the Atmosphere" Presented in National Workshop on

Atmospheric Chemistry, Indian Institute of Tropical Meteorology, Pune, by S.K. Chopkar,

IRRA Scientist Group, India.

CERTIFICATE FROM IITM, PUNE CONFERENCE ON ATMOSPHERIC CHEMISTRY

NATIONAL WORKSHOP ON ATMOSPHERIC CHEMISTRY
(NWAC - 99)
Indian Institute of Tropical Meteorology, Pune

CERTIFICATE OF PARTICIPATION

This is to certify that Dr. Shivkumar Chopkar *from* ARRS, Chandrapur *has participated in the "National Workshop on Atmospheric Chemistry", held at IITM, Pune from 12 to 14 October 1999.*

(Dr. P.C.S. Devara)
Co-ordinator, NWAC-99

DR. P. C. S. DEVARA
Co-ordinator, NWAC-

18.2 International Conference on Environment Chemistry, at Shri Venkatesh University, Tripathi by S.K. Chopkar, IRRA Scientist Group, India.

Research Paper "Rainmaking Technology by Laser System Initiations Endothermic Reactions, condensations, Precipitations, Raindrops which act as Natural seeding process, Enhancing for Rainfall in the Atmosphere" Presented in International Conference on Environment Chemistry, at Shri Venkatesh University, Tripathi by S.K. Chopkar, IRRA Scientist Group, India.

(Flow Chart for presentation in VENKETESH UNIVERSITY TRIPUTI CONFERENCE)

"Innovative Rainmaking Technology by Laser System initiating Endothermic Reactions

similar to Natural Lightning Phenomena for Rainmaking in the Atmosphere"

**After Lightning heavy rainfall occurs → *Lightning Phenomena demonstrated by Laser system in Laboratory cloud chamber → * Bonds of N_2 & O_2 breaks, at a high temp. → *Excited [N] & [O] form → Excite N* and O* are unstable→*

**React to each other → *Form NO & O_3 → *Endothermic reactions → *Lot of heat energy Required for these Reactions →* which is absorbs from the surrounding atmospheric clouds →*Condensations take place → *Precipitation form with tiny water drops → * "Laser induce condensation & water drops formations in Laboratory cloud chamber". →These tiny water drops in the atmosphere*

**Due to wind forces, such as acceleration & turbulence forces in the atmosphere→ These tiny water drops act as → *Natural seeding process to form, another set of raindrops → *Chain process occurs in the atmosphere → *Heavy rainfall occurs by rainmaking technology, Similar to lightning phenomena → *Practical results shows in the Laboratory "Laser induce condensation and water drops formed in laboratory clouds chambers" → Also, *Practical results show in the atmosphere*

"Laser induce condensation and water drops formed in the atmosphere" →

** Artificial Rainmaking in the way of natural lighting phenomena by the Laser system from the Ground, in the atmosphere. →Innovative Rainmaking Technology,*

It's scientifically and practically proven, in Laboratory cloud chambers as well as in the atmosphere

18.3 In International Environment Conference, University of Calcutta-700009, India by D.K. Chakrabarty, IRRA Scientist Group, India.

Research Paper Presented "Femto-second Terawatt Laser system to produce Artificial Rain" in International Environment Conference, University of Calcutta-700009, India by D.K. Chakrabarty, IRRA Scientist Group, India.

18.4 comparison statement with calculation on energy and reactions

"Comparison of Energy statement calculated byDr. S.K. Chopkar"

**Heat energy generated /evolved from said lightning channel in the upper atmosphere by Lightning → $4.22x10^4$ Kcal.*

**Heat energy utilized for breaking the bonds of N_2 & O_2 in the said lightning channel → $4.16x10^4$ Kcal*

**Means, lightning energy is completely utilized for breaking the bonds.*

**Heat energy absorbed by endothermic reactions from surrounding atmospheric clouds → $8.37x10^3$ Kcal in the said lightning channel.*

**Heat energy abstracted from water vapors in the said lightning channel for dew point to form water drops → $1.58x10^3$ kcal.*

** $8.37x10^3$ Kcal is much more than $1.58x10^3$ kcal. Means, endothermic reactions will affect not only inside of the said lightning channel, but also more area of the surrounding atmosphere covered.*

Above calculation was published in "Indian Journal of Radio Space Physics" in Vol 22, (April 1993 issue), as" Effect of endothermic reactions associated with lightning on atmospheric chemistry" by S.K. Chopkar

"Calculation and comparison statement of dissociation and ionization by prof. D.K. Chakrabarty"

**For dissociation process, 1 molecule of N_2 and 1 molecule of O_2 , energy required → 2.2 joules.*

**But for ionization process, 1 molecule of N_2 and 1 molecule of O_2 , energy required → 4.4 joules.*

**As per above calculation, first dissociation @2.2 joules take place, if energy balance remains there, ionization takes place @ 4.4 joules.*

As per above calculation, each and every laser shooting in laboratory cloud chamber, NO and O_3 formation are observed by laser filament. But

N_2^+ and O_2^+ are not observed or measured by laser shooting in the laboratory cloud chamber.

"Laser induce condensation and water drops formation" in the laboratory cloud chamber. Who is responsible for this condensation; an only endothermic reaction? in dissociation process is responsible.

Above result has been published in "The International Journal of Meteorology" www.ijmet.org,Volume *35, No.355, Nov.2010 as "Artificial Rainmaking by laser system".*

J. Kasparian and TML Scientist Group have dropped their ionization theory and have agreed with NO and O_3 formation by Laser.

18.5 "Research Activities & Achievement for Innovative Rainmaking Technology" By IRRA Scientist Group, India

"Achievements"

Research Paper Presented in National and International Conference By IRRA Scientist Group

**Research paper "Novel Technology for Artificial Rainmaking by Laser system in a way of Natural Lightning Phenomena" presented in International Environment Conference @ Dakhla, Morocco country, sept.2018 by S. K. Chopkar, IRRA Scientist Group, India.*

**Research Paper "Rainmaking Technology by Laser System Initiations Endothermic Reactions, condensations, Precipitations, Raindrops which act as Natural seeding process,*

Enhancing for Rainfall in the Atmosphere" Presented in National Workshop on Atmospheric Chemistry, Indian Institute of Tropical Meteorology, Pune, by S.K. Chopkar, IRRA Scientist Group, India

**Research Paper "Femtosecond Terawatt Laser system to produce Artificial Rain" in International Environment Conference, University of Calcutta-700009, India by D.K. Chakrabarty, IRRA Scientist Group, India.*

**Research Paper "Rainmaking Technology by Laser System Initiations Endothermic Reactions, condensations, Precipitations, Raindrops which act as Natural seeding process, Enhancing for Rainfall in the*

Atmosphere" Presented in International Conference on Environment Chemistry, at Shri Venkatesh University, Triputi, (A.P.)

Award achievement to IRRA Scientist Group, India

**IRRA Scientist Group, has been awarded International Scientist Outstanding Award, (2019) by VDGOOD Professional Association, Chennai, India.*

Publish Soft ware

**Please visit our YouTube Channel search as" Novel Technology for Artificial Rainmaking' .Click here :- https://youtu. be/UYfuH9fAmUs*

**Please search on Google as S.K. Chopkar or "Rain Enhancementby Endothermic Reactions "or visit our web site http://www.irraindia.org*

National & International Patent Certificate and Documents.

** International Australian Government, IP Australia Certificate of Grants Innovation Patentnumber: 2020101897(2021).*

**International Publication Patent No: -WO/2008/062441(2007) and RevisedInternational Application No.PCT/IN2017/000105 (2017).*

**Indian Patent Number: - 238000 (2007) and Revised Indian Patent File Number*

201721008920(201

19.0: Registration Certificate For IRRA

GOVERNMENT OF INDIA MINISTRY OF CORPORATE AFFAIRS

Central Registration Centre CERTIFICATE OF INCORPORATION

[Pursuant to sub-section (2) of section 7 and sub-section (1) of section 8 of the Companies Act, 2013 (18 of 2013) and rule 18 of the Companies (Incorporation)

Rules, 2014]

I hereby certify that INNOVATIVE RAINMAKING RESEARCH ASSOCIATION is incorporated on this Fifth day of February Two thousand twenty under the Companies Act, 2013 (18 of 2013) and that the company is limited by shares.

The Corporate Identity Number of the company is U73100MH2020NPL337195.The Permanent Account Number (PAN) of the company is AAFCI5742P

**The Tax Deduction and Collection Account Number (TAN) of the company is NGPI02173D*

**Given under my hand at Manesar this Fifth day of February Two thousand twenty, (15/02/2020).*

Digital Signature Certificate

Sd/-

M MOHAN ASST. REGISTRAR OFCOMPANIES

For and on behalf of the Jurisdictional Registrar of Companies Registrar ofCompanies Central Registration Centre

Disclaimer: This certificate only evidences incorporation of the company on the basis of documents and declarations of the applicant(s). This certificate is neither a license nor permission to conduct business or solicit deposits or funds from public. Permission of sector regulator is necessary wherever required.

Registration status and other details of the company can be verified on www.mca.gov.in

Mailing Address as per record available in Registrar of Companies office: INNOVATIVE RAINMAKING RESEARCH ASSOCIATION

C/O VIJAY MADHUKARRAO KHODE,

EKNATH TEMPLE, SUDAMPURI, WARDHA, Maharashtra, India, 442001

20.0: International Scientists Award

International Scientist Award Ceremony Program Photos

Demonstration for Innovative Rainmaking Technology for Artificial Rain in large scale by creating multiple artificial lightning by multiple high-power Laser which initiates endothermic reactions, similar as nature's lightning phenomenon, onboard Aircraft in the upper atmosphere"

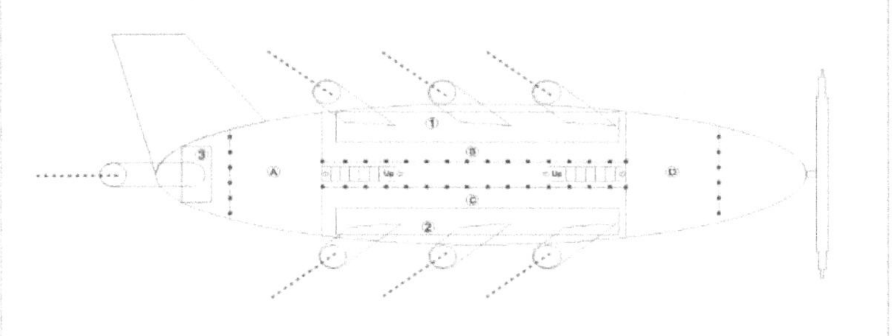

Fig. "Position of Laser system in Aircraft"

**** In above figure on first floor, height 9', shown parts (1), (2) and (3) are high power Laser release system which creates multiple artificial Lightning in the upper atmospheric clouds directly by Laser system ......***

****In above figure, on second floor, height 7' 6", shown parts (A), (B), (C) and (D) with stair case are high power generation system for high power Laser system, with railing for open space as shown square dots ....***

"IRRA Scientist Group propose laser system of specification: 10^{12}watt, 800nm, 500mJ, 120fs and 10Hz for this research project for demonstration with multiple Laser creates rain in large scale.

Please visit our YouTube Channel and search "Novel Technology for Artificial Rainmaking". Click here: - https://youtu.be/UYfuH9fAmUs

More details can be found in Google Search: as "Artificial rainmaking by endothermic reactions" or S.K. Chopkar or on the website: http://www.irraindia.org

Prof D.K. Chakarbarty
(Scientific Director of IRRA Scientists)
E- dkchakrabarty@rediffmail.com
M+919327020993

Dr.S.K. Chopkar
(Director of IRRA Scientists)
M +91 9420445108,
skc.arr@rediffmail.com

Above are two pillars of our organization "Innovative Rainmaking Research Association (IRRA Scientists) India. We thank them on behalf of IRRA Scientists Group India for making such a wonderful platform to work on Innovative Rainmaking Technology used for Artificial rainmaking by Laser system which initiates endothermic reactions, similar as nature's lightning phenomenon, onboard Aircraft in the atmosphere

www.ingramcontent.com/pod-product-compliance
Ingram Content Group UK Ltd.
Pitfield, Milton Keynes, MK11 3LW, UK
UKHW062258290726
14090UKWH00017B/759